The Theory of Credit Contracts

Christian Prem

The Theory of Credit Contracts

With a Focus on Group Lending

Springer Gabler

Christian Prem
Universität Regensburg
Universitätsstraße 31
Regensburg, Germany

University of Regensburg
Faculty of Business, Economics and Management Information Systems
Doctoral Thesis in Economics
Supervised by Prof. Dr. Lutz Arnold and Prof. Dr. Andreas Roider
Date of Disputation: 06. December 2018

ISBN 978-3-658-29361-1 ISBN 978-3-658-29362-8 (eBook)
https://doi.org/10.1007/978-3-658-29362-8

This Springer Gabler imprint is published by the registered company Springer Fachmedien Wiesbaden GmbH part of Springer Nature.
The registered company address is: Abraham-Lincoln-Str. 46, 65189 Wiesbaden, Germany

Preface

The desire of this monograph is to improve peoples' lives by improving our understanding of credit relationships. In keeping with the Lewinian mantra that there is nothing so practical as good theory[1] it hones the theoretic side of the field, it *improves* and *enriches* the theory of credit contracts.

In part I. *improvement of the existing theory* is achieved by a new approach that carefully decomposes the field into reasonably isolated subproblems which can then be researched quite independently (chapters 2, 3, 4). This unconventional modular approach to the theory of credit contracts is opposed to the rather integral and monolithic research practice that predominates in the literature. It can be seen as a complete restructuring of the theoretic landscape and yields many benefits: It provides a framework to clearly arrange existing knowledge, it enables specialisation and cooperation, and it creates transparency by neatly unravelling the complex interplay between different aspects of lending theory. For instance contract effects and market effects are explicitly separated. In addition to introducing this new modular approach, part I. bestows great care on providing precise definitions of the concepts used in credit contract theory and chapter 5 promotes working towards a standardisation of the field to increase research efficiency.

In part II. enrichment comes about by a whole bunch of *new results*. In order to produce clear-cut conclusions, the structure developed in part I. is applied and the modules are researched individually. In chapter 6 the idea that joint liability mitigates adverse selection is scrutinised and proven wrong in several cases. Chapter 7 provides a very general proof for the risk shift between lender and borrower that is induced by using collateral and it examines the effects of this result on other modules of the theory. Subsequent chapters concentrate on the improvement of repayment games by reinvestigating existing and inventing new payback-enforcement schemes. A special feature is the uniformity of the settings in chapters 10, 11, 12 which makes different payback-enforcement schemes easily comparable and gives practitioners the chance to choose the context dependent best one (see figure 13.2). Chapter 9 introduces a Mathematica-Code that can solve and analyse repayment games automatically. In chapters 10, 11, 12 this tool is put to use, to effectively research many payback-enforcement schemes. In 10 I present a new model that reproduces the perverse effect of joint liability with a minimum of assumptions and

1 See [Wei03, p. 459].

I demonstrate a simple but effective countermeasure, namely adaptive punishment. The applicability of this punishment strategy in the real world is discussed, too. In chapter 11 a new model is introduced which is explicitly tailored to shedding light on the effects of social sanctions – in joint liability contracts with two borrowers and with three borrowers, as well as in individual liability contexts. Finally, chapter 12 reviews the quality of cross-reporting mechanisms presented in the literature and invents a new and better one that features zero equilibrium punishment and only unique equilibria in every possible state of the world. This robustness to different states makes the result independent of the actual probability distribution, thus the application of this mechanism is safe for any real world distribution. As a by-product the model provides a new formal explanation for the trend to keep banking meetings despite the change back to individual liability contracts by microfinance institutions.

Altogether this dissertation features new innovations on several levels. On a conceptual level the modularisation of lending theory, on a formal level the standardisation of lending theory and precise definitions, on a technical level the development of an algorithm to solve repayment games automatically, and with regards to content several new theoretic results on the effects and the performance of various credit contracts. By pointing at numerous open questions and problems, this monograph also motivates future research projects in the field of credit contract theory, in order to extend the boundary of our knowledge even further.

Although it is me who wrote this thesis, it is a whole crowd of people that put me in the position to do that. Most directly my supervisors Prof. Dr. Lutz Arnold and Prof. Dr. Andreas Roider who I could always turn to with my questions. I was also fortunate to be surrounded by colleagues of great heartiness that often shared their knowledge as trained economists whilst I was working myself into the field (as a trained mathematician). In particular I had long discussions with Marina Markheim, a specialist in Microfinance. Of course, my knowledge and skills have been developed with the assistance of many teachers at school and professors at university and also students asking clever questions in my own lectures. I view this thesis as a culmination of all their efforts and enthusiasm. Finally, I enjoy the company of some close companions who acted as supporters, who cared for me during hardships and who constitute an invaluable wealth. Thank you my friends!

Regensburg, April 2019 Christian Prem

CONTENT HIGHLIGHTS

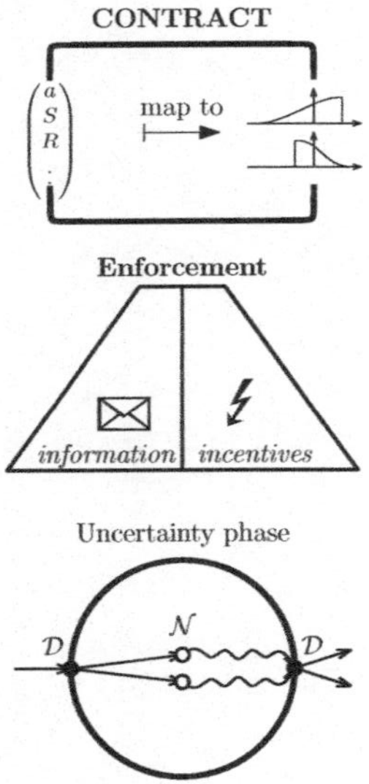


```
(* GAME DEFINITION — Individual liability *)

(* NUMBER OF PLAYERS *)
n=2;

(* ACTIONSET for each player *)
actionSet[1]={
{"0R",0,""},
{"1R",−R,""},
{"2R",−2R,""}};
actionSet[2]=actionSet [1];

(* PUNISHMENT FUNCTION: borrower to deadweight flow *)
punishmentFunction[abilityCombination_,actionCombination_,
        playerKey_]:=Refine[Piecewise[{
{−M,      −fnGetActionRepayment[actionCombination[[playerKey
        ]]]<R (* punishment if default ... *)
&& −Plus@@(fnGetActionRepayment/@actionCombination)< 2 R (*
        ... and peer did not kindly pay for him *)},
{0,            True}(* No punishment otherwise *)
}]];

(* REWARD FUNCTION: lender to borrower flow *)
rewardFunction[abilityCombination_,actionCombination_,playerKey_
        ]:=0;

(* SANCTIONS FUNCTION: borrower to deadweight flow *)
sanctionFunction[abilityCombination_,actionCombination_,playerKey_
        ]:=0;

(* ASSEMBLING PAYOFFLAYERS (apart from repayment layer) *)
payoffLayers[abilityCombination_,actionCombination_,playerNr_]:=+
        punishmentFunction[abilityCombination,actionCombination,
        playerNr] + rewardFunction[abilityCombination,
        actionCombination,playerNr]+ sanctionFunction[
        abilityCombination,actionCombination,playerNr];

(* ASSUMPTIONS *)
$Assumptions=True; (* Clearing list of global assumptions *)
$Assumptions=And[
R> 0, (* sign gross interest *)
M>R]; (* intensity punishment *)
(* Use == for comparisons! No contradictory assumptions! No
        ambiguous assumptions like <= ! *)
```

Adverse Selection & Joint Liability:
Risk-averse and low income borrowers squeezed out

Risk Shift Effect of Collateral:
Most general proof and possible consequences

Perverse Effect of Joint Liability:
Maximum repayment through adaptive punishment strategy

Mechanics of Social Sanctions:
New cleaner model, motivation & coordination effects, extension to groups of three and individual liability

Cross-Reporting Schemes:
New mechanism featuring zero equilibrium punishment

NEW FRAMEWORK

The field of lending theory structured into layers and modules

NEW TOOL

A new Mathematica-Algorithm to solve and analyse repayment games

NEW RESULTS

Collateral risk shift. Adverse selection under JL. Improved repayment games.

Contents

List of Illustrations

List of Infoboxes

List of Literature Analyses

Part I

General Topics

1 Start: The meaning of credit

Is research into credit contracts valuable to people? Or is it just brain callisthenics in the ivory tower of science? The first chapter of this book will

- define *what* credit is,
- demonstrate *why* credit is of enormous social relevance and
- explain *how* this valuable tool releases its incredible potential.

Credits change peoples lives! It is of vital importance to understand them well.

1.1 Credit – a disambiguation

We start out with the fundamental concept of this dissertation – the definition of the term 'credit'.[2]

Definition (Credit). We speak of *credit* if one party, named creditor or lender, transfers money or resources to another party, named debtor or borrower, where the second party does not reimburse the first party immediately but instead promises to return something of value at a later date.

First, let's pinpoint the core. The constitutive feature of credit is the surrender of something valuable *without immediate consideration*, but with a *compensation promise*. Or as Hermann Feifel puts it: "The constitutive feature of credit is the lag between service and return service." [Fei59, p. 74][3] This time lag gives rise to a so-called *uncertainty phase* whose analysis is at the heart of credit contract theory and modelling it reasonably provides a challenge (see p. 23).

2 The following definition is based on the definition of credit in "Gabler Bank-Lexikon" [GGH+13, p. 873, term "Kredit"]. The words "capital" and "purchasing power" have been replaced by the more general expression "resources".

3 Translated from the German original: "Konstituierendes Merkmal einer Kreditbeziehung ist eine Zeitdifferenz, die auftritt zwischen dem Zeitpunkt der Leistung und dem Zeitpunkt der Gegenleistung."

C. Prem, *The Theory of Credit Contracts*,
https://doi.org/10.1007/978-3-658-29362-8_1

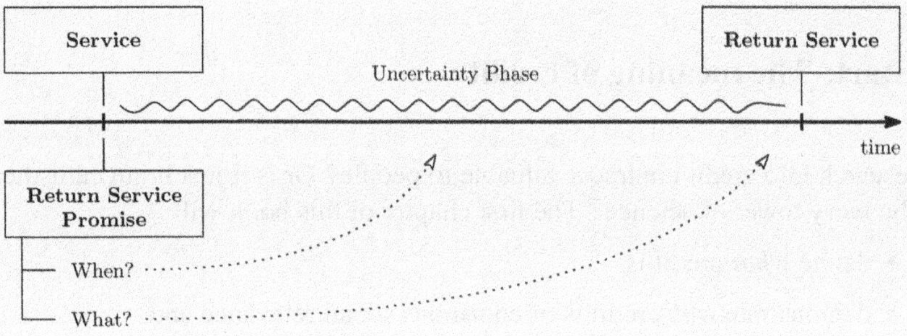

1.1 Constitutive feature of credit
The constitutive feature of credit is the time lag between service and return service.

Second, notice that the service and the promised return service are not necessarily of equal value! Often the compensation promise includes the principal plus an additional amount called *interest*. Given a interest-free loan with no inflationary adjustment, the promised return service may also be less valuable than the initial service.

Third, note that despite the fact that definitions of the term ‘credit’ often restrict their focus on the surrender of money, credit settings are not at all restricted to bank loans! The concept ‘credit’ is applicable to all situations in which some required compensation is not carried out immediately. It designedly applies to a whole variety of situations and contract types. Consequently, the theory of credit relationships is applicable to much more than just bank loans – which makes it all the more valuable.

Q 1.1 Definition of credit in other fields

Notice that in other contexts the term "credit" may be defined differently. E.g. the jurisprudential meaning differs from its economic meaning, as Andrey Shatelyuk explains in [Sha12, p. 23–26]. Referring to German legislation, he explicates: "The term credit as it is used in economic sciences is too broad for legal implementation."[a] This difference becomes comprehensible if you consider the different purposes in both fields. Economists want to circumscribe the general principle, the conceptual core of credits. The more general the definition, the greater the number of real world situations that can be subsumed. Thereby, the theory becomes applicable to a broader range of examples. Economists zoom out and catch the essence. In contrast attorneys and judges need a definition which is rather explicit, such that decisions in concrete law cases can be based upon it. Their needs require to zoom in and codify the details.

a Translated from the German original: "Der Kreditbegriff in der Wirtschaftswissenschaft ist zu breit für die rechtliche Umsetzung."

A final note on the wording in this dissertation. I will often treat the following pairs as synonyms: creditor/lender, debtor/borrower, credit contract theory/lending theory, credit/loan/debt. There might, however, be a slight semantic difference. A lender may be interpreted as a *potential* creditor, someone who in general lends money to others, whilst a creditor is someone that is in fact engaged in a credit contract. Likewise 'credit contract theory' may refer explicitly to the functionality of credit contracts, whilst the term 'lending theory' is a bit broader and also includes the process of contracting.

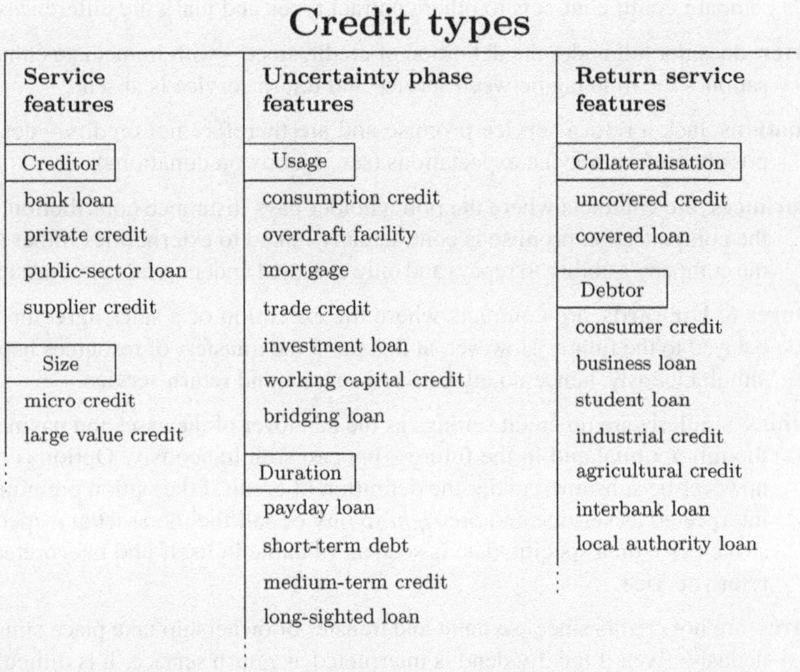

1.2 Classification of credit types
The dazzling array of different types of credits can be sorted along various different criteria. The illustration provides some examples. Of course there exist even more features that can distinguish different types of credits like the type of resources lent, repayment frequency, fixed vs. non-fixed repayments etc.

Let's proceed from the definition of credit to examples. Whether a customer buys her new ball gown by instalments or whether they agree on cash before delivery, both alternatives represent a credit setting. In the former case the role of the debtor is taken by the customer whilst the salesperson represents the creditor; in the later case roles are vice versa. Employees regularly appear as creditors when they perform their contractual job performance before receiving their pay. Just as tenants if they pay their rent up front.[4] Note that *lending* theory in particular applies to *saving* – it is just a question of who plays which role. Figure 1.2 shows how the bulk of different credit settings can be structured.

Let's compare credit contracts to other contract types and mark the differences.

Barter does not fall under the definition of credit, since – with immediate compensation – the time lag between service and return service is absent.

Donations lack a return service promise and are therefore not credits – despite possible return service expectations (see infobox on donations).

Insurances are contracts where the policyholder pays insurance contribution, but the compensation promise is contractually linked to externalities (other than the company's ability to repay) and only executed under certain circumstances.

Futures & Forwards are contracts where the execution of a sales agreement is delayed to the future. However, at that point the transfers of resources happen simultaneously, hence no lag between service and return service.

Options similarly are no credit settings as the handover of the asset and payment – though optional and in the future – happen simultaneously. Options could however be subsumed under the definition of credit if the option premium is interpreted as service and *the right* to buy or sell the the asset at a specific strike price on a specific date is seen as valuable in itself and interpreted as return service.

Shares are not credits since payment and transfer of ownership take place simultaneously. Even if the dividend is interpreted as return service, it is difficult to interpret shares as credit setting, since the amount of dividends is decided at the annual shareholder's meeting, and there is no specific return service promise at the time of acquisition.

4 The service here is the assignment for beneficial use of the object, e.g. for a month. And this service is finished not until the end of the month. So there is a delay between the service and the completion of the return service.

Q 1.2 Reciprocity of donations
Notice that social conventions may require some kind of return service even in the case of donations. Nevertheless, reciprocity with a donor *expecting* some return service is not of the same quality as reciprocity with a receiver that *promises* a return service. So, even though there may be a compensation expectation, the decisive criteria for a credit situation is a compensation promise.

Notice that the definition of credit becomes narrower if the 'When' and 'What' of the return service promise is required to be very specific. E.g. if you require the duration of the uncertainty phase to be fixed in the return service promise, credits of flexible duration may fall out of the definition. Conversely, if you allow the promise to be very unspecific and say it is sufficient to pay back something at some time, then even the dividends of shares may count as return service promise. In this monograph the stipulated return service has a fixed value, denoted by R, and is delivered at a fixed point in time.[5]

1.2 Credit – an enhanced form of reciprocity

Having defined the meaning of credit, we can now begin to examine its relevance. Reciprocity, the old give and take, the social rule saying people reciprocate what they have received from others, is a basic principle of human interaction. M. Simkovic points out that credits can be interpreted as an enhanced form of reciprocity: "Credit markets help coordinate activities by expanding the concept of reciprocity – "I will do something for you now if you do something for me later" – to vast numbers of individuals who are unrelated and do not know each other." [Ame16, p. 1] How advanced this form is, becomes clear when we carefully distinguish different levels of reciprocity.

No reciprocity – which is of course barely imaginable – is the lower bound of the staircase in figure 1.3. Without reciprocity no exchange of resource and hence no cooperation is possible.

Immediate reciprocity refers to barter trade without time lag. All obligations are satisfied immediately, there is no phase of uncertainty. Cooperation is possible, but only to a very restricted extent and only if the wishes of both trading partners do match accurately.

5 The concrete duration of the uncertainty phase will not matter as there will be no dynamic modelling of time in this monograph.

Deferred reciprocity provides a drastic extension of the complexity of cooperation by allowing time lags. The parties concerned can defer the fulfilment of their duties. Or as Simkovic puts it: "I like to think of credit as "time travel for reciprocity."" [Ame16, p. 1] However, the uncertainty phase between service and return service requires trust and thus restricts this form of deferred reciprocity to basically intimates.

Formalised and enforceable deferred reciprocity, eventually, yields another enhancement of social cooperation by enabling the sharing of resources between unacquainted individuals. It tremendously increases the number of people that can cooperate with each other. This is what formalized loan contracts do. They replace trust by enforceable contracts and thus render deferred reciprocity between strangers possible. Eventually, the credit markets mentioned by Simkovic bring together potential lenders and borrowers.

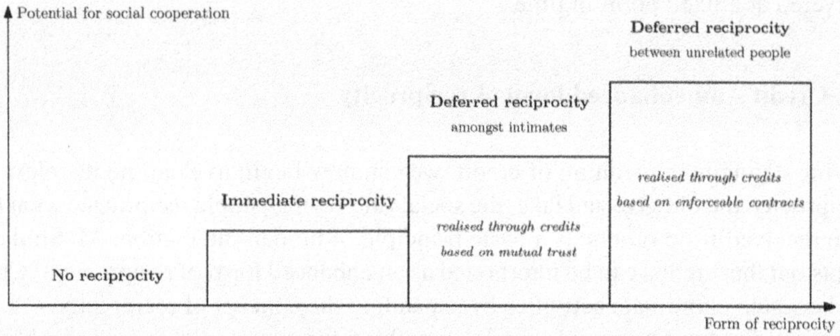

1.3 Levels of reciprocity
Credits expand the possibilities for social cooperation.

1.3 Classes of reciprocity in sociology

Sociologists distinguish several classes of reciprocity. Most notably they differentiate between *direct reciprocity* and *generalised reciprocity*. Direct reciprocity is based on direct relationships. The parties mutually perform approximately equivalent services in a clearly nameable trade. An example is barter. Generalised reciprocity refers to services that are provided with the hope for some form of compensation but without a *direct* compensation promise. Examples are groups where each member contributes and takes but services cannot be decomposed into clearly marked-off pairs of service and return service, e.g. families or sports clubs. For details see [Ste11, p. 29, 33]. Credit relationships are part of the direct reciprocity class. If the term *reciprocity* is used in this monograph without further indication, it always refers to direct reciprocity.

1.3 Credit – the tool for massive multilateral cooperation

So far it has become clear that credit contracts create new possibilities for *bilateral* cooperation between unrelated people. To tap the full potential of credit, however, several bilateral contracts can be combined to enable massive *multilateral* cooperation! We can distinguish two basic arrangements of credit contracts: Sequential cooperation and simultaneous cooperation (see figure 1.4). The former renders possible exceptionally long-running projects, whilst the later enables particularly extensive projects or exceedingly risky ones.

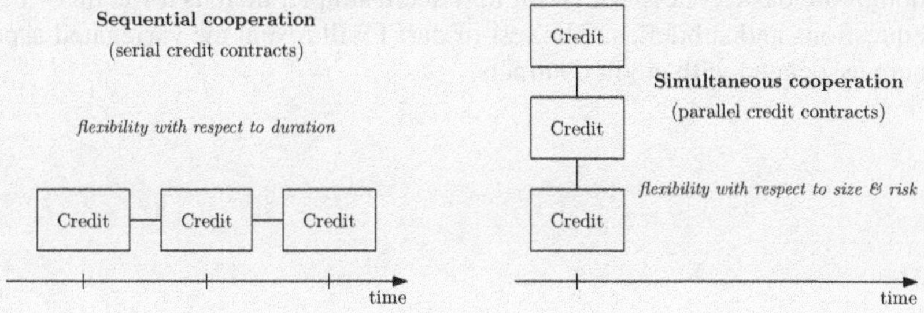

1.4 Credit flexibility with respect to duration and size
Serial and parallel combinations of credit contracts lead to flexibility in duration and flexibility in size or risk, respectively.

Flexibility with respect to duration is achieved when credits are rolled over, i.e. when many people organise their credits *sequentially*. Extremely long-term debts become possible – they can even span generations. Transformation of size happens when people synchronise their deferred reciprocity agreements and many creditors finance one debtor. Large projects are realisable.[6] A direct consequence of the flexibility in size is the flexibility in risk. If a loan is split into several small pieces, the risk of the debtor being unable to repay can be distributed amongst many creditors. Even though one creditor could finance a debtor on its own, it makes sense to spread the burden on many shoulders. This puts societies in a situation where venturing risky projects becomes bearable.

If multilateral cooperation is processed by intermediaries like banks, the associated transformations of duration, size and risk must be carefully managed! Otherwise,

6 By the way, it also works the other way round. If you are a very wealthy person with many idle funds, maybe no single debtor is in need of such big loan. Then flexibility in size allows to divide the loan and distribute the resources amongst a number of borrowers.

bank runs, collapses of financial institutes even culminating in economic crises are possible consequences. Massive multilateral cooperation provides incredible chances, but the increased complexity exposes society to systemic risks. This is just another reason why we should understand lending theory well.

The variety of ways we can make use of credit is impressive. We help each other in building houses, roads and infrastructure, we put our children through college, we finance research and the development of new products, we save money for hardships and retirement. Credits are omnipresent, they are essential for our living together, for our prosperity, for our lives. At heart, our society is unthinkable without credit! The goal of lending theory is to deepen our understanding of these credit relationships. Although the basic concept of credit may seem simple, there is a wealth of issues and questions and subtleties. The rest of part I will reveal the variegated aspects that are associated with credit contracts.

2 Structure: The structure of credit contract theory

This chapter and the rest of part I are devoted to an in depth analysis of the structure of lending theory. A description of the basic structure of credit contracts has been given in chapter 1 already (see service, uncertainty phase and return service on p. 3). We now focus on gaining a clear conceptual understanding of the various aspects and details of lending theory. The goal is to partition the whole subject into reasonable sub-problems, to carve out its individual 'organs' and arrive at something like an 'anatomy' of lending theory.

2.1 Purpose: Complexity control

Why is classification and systematisation so important to master a subject? Why do anatomists partition the human body into organs and subsystems? Why do software engineers build modularly structured software? Why does the level of detail reduce when you zoom out of an online map? The reason: Complexity control! Due to the restrictions on human intelligence, it is necessary to subdivide complex structures into manageable substructures.

Although structuring the subject is an effort that will not deliver any new results in the first place, it is a necessary and worthwhile first step and the fertile soil for research quality and efficiency. It is a prerequisite to keep sight of the complexities and interdependencies in our subject.

C. Prem, *The Theory of Credit Contracts*,
https://doi.org/10.1007/978-3-658-29362-8_2

What are the advantages that come with a modular approach to lending theory?

- The most obvious one: We gain cognitive facilitation. A good structure brings orientation, it improves the understanding of connections and interdependencies, it provides didactic advantages. In a world where the cutting edge of research is shifted outwards ever rapidly, it is crucial to speed up the process of sharing expertise.
- Even more importantly, modularisation enables specialisation. In credit contract theory we encounter a mix of game theory and probability theory. Specialists can design individual modules more realistically without increasing overall complexity, because the interior of a module remains a black box to all other modules. Reasonably demarcated substructures can therefore ease the conflict between realism and simplification of theoretic models.
- Finally, modular research is more comprehensive. By recycling and recombining modules we cover many different cases and situations very easily – without the need to build and calculate an entirely new model each time. (For more on the separation and combination of modules, see chapter 4.)

To set up a modular approach and harvest these benefits, we have to demarcate reasonably isolated sub-problems of credit contract theory. *Reasonably isolated* means that the substructures can be researched relatively independently. The different parts will necessarily interact with each other but the modules must be defined in a way such that the interfaces between these modules become as simple as possible. The quality of demarcation is measured by the simplicity of the connecting interfaces and hence the possibility to treat modules individually.

2.2 Literature: Existing structures and their weaknesses

In the lending literature, it is quite common to build big monolithic lending models without putting a particular emphasis on isolating subproblems and reducing complexity. Unfortunately – as a statement J. M. Keynes from 1936 proves – this seems to be a quite persistent problem of economics. "Too large a proportion of recent "mathematical" economics are merely concoctions, ... which allow the author to lose sight of the complexities and interdependencies of the real world in a maze of pretentious and unhelpful symbols." [Key36, p. 297] The attempt, however, has been made to get some structure into lending theory. How do other writers organise lending theory? Let's have a quick look at some publications that treat credit contracts fairly comprehensively to see how they structure the field.

2.2.1 Existing attempts to structure lending theory

2.1 How other authors structure the field of lending theory

[FR08]	**"Microeconomics of Banking"**
4	"The Lender-Borrower Relationship"
4.1	"Why Risk Sharing Does Not Explain All the Features of Bank Loans"
4.2	"Costly State Verification"
4.3	"Incentives to Repay"
4.4	"Moral Hazard"
4.5	"The Incomplete Contract Approach"
4.6	"Collateral as a Device of Screening Heterogeneous Borrowers"
[Han01]	**"The Structure of Credit Relationships"**
2	"Essential Components of Loan Commitments in Financial Theory"
2.1	"Interest rate"
2.2	"Securities"
2.3	"Credit volume"
2.4	"Principal bank status"
[GG99]	**"The economics of lending with joint liability: theory and practice"**
2	"Theories of joint-liability contracts"
2.1	"Adverse selection"
2.2	"Moral hazard"
2.3	"Costly state verification"
2.4	"Enforcement"

In [FR08] – a comprehensive textbook about the microeconomics of banking in its second edition – chapter 4 is dedicated to the lender-borrower relationship (see above). In [Han01] chapter 2 deals with central aspects of credits in finance theory (see above).[7] So far, no common framework is identifiable. A recurring structure, however, can be found if we turn to the microfinance literature. Authors seem to brachiate through lending theory along information asymmetries and incentive problems. The often cited overview paper about the economics of lending with joint liability by Ghatak and Guinnane partitions its theory section this way (see above). Fischer and Ghatak in 2011 ([FG11, p. 63]) refer to this structure as the four major 'problems' facing lenders, namely: screening, monitoring, auditing, enforcement.[8] Also chapter 4 in "The Economics of Microfinance" ([AM10]) by Armendáriz and Morduch proceeds in a similar way. And [Zha08, p. 146 ff.] is just another example using this structure. Simtowe and Zeller (2006) contrast these problems with possible countermeasures provided by group lending, see [SZ06, p. 3].

7 Titles translated from the German original.

8 These terms, of course, rather refer to solutions than problems. For example screening is one possibility to counter adverse selection.

2.2.2 Weaknesses of existing structures

As we have just seen, the attempt has been made to structure the theory of credit contracts. However, there is still room for improvements. Here are some problems that may arise.

Fitting credit contract theory into a problems-versus-solutions structure like in [SZ06, p. 3] is tempting. But the first issue is that it is not at all well-defined what should be considered a 'problem'. The classification of some phenomena as problems is a rather subjective matter. Second, solutions and problems do not form an unambiguous assignment as some 'solutions' target several 'problems' and a 'problem' can have several different 'solutions'. Moreover, some so-called 'solutions' can create new 'problems' again.[9] Thus a closer look reveals that this approach does not produce nicely separated substructures.

Also a rather chronologic approach exhibits weaknesses. A frequent structure used is to sort topics into ex ante topics and ex post topics. According to M. W. Jenkins *ex ante* and *ex post* are defined as follows: "The former refers to incentive problems affecting agents' behaviour before the realization of uncertainty (...), while the later affects agents' behaviour after the realization of uncertainty (...)." [Jen09, p. 90][10] The chronological distinction between ex ante and ex post topics has the problem that the uncertainty that is used as a fixpoint may not be unique. Another problem of chronologic approaches is that the timelines of lending models have a certain degree of flexibility (we will analyse timelines in chapter 5, see p. 109). Altogether, cutting the timeline in two pieces is simple but can be problematic and does not necessarily produce reasonably marked-off modules.

The idea to contrast informational aspects and incentive aspects like [FG11] do with their topics screening, monitoring, auditing, plus enforcement ignores the fact that effective contract enforcement always requires a combination of observational *and* sanctioning capabilities. Thereby they separate two things that inevitably belong together (see figure 2.4 on p. 24). Since monitoring and auditing only take effect with adequate incentives (otherwise the gained information would be in vain) a separate treatment of both does not seem opportune. Again, the isolation of different aspects of lending theory does not seem to be optimised for modular researchability.

9 E.g. collateral can be a solution to adverse selection, see [Bes85]. However, depending on how it is used, high collateral requirements can also lead to adverse selection, see [SW81, p. 402]. So does collateral belong to the solution box or the problem box of lending theory?

10 According to [AM10, p. 50] "the term ex post refers to difficulties that emerge after the loan is made and the borrower has invested." Setting the boundary between ex ante and ex post to the point of *investment* would term all hidden actions during the runtime of an investment project as ex post. Yet problems like low effort (p. 47) are typically counted among ex ante problems. Curiously, on page 48, for the definition of ex ante moral hazard, they refer to the *realization* of project returns.

2.3 Modularisation: The PRE-layer-system

Having discussed the shortcomings of existing structures, this section presents a new proposal. It features a careful deconstruction of lending theory into different layers, namely **p**roduct layer and contract layer (which further decomposes into **r**eciprocity and **e**nforcement).

Why do we introduce layers? It is a like zooming in and out a map: Too close and you lose the big picture, too far and you miss the details. The art of designing good abstractions means finding the right distance to your object of investigation. The layers will enable looking at lending theory with different levels of granularity.

In other fields like mathematics or computer science forming abstractions is daily business.[11] But in economics, too, we can zoom in and out. We can zoom from individual actions in a particular situation (e.g. a repayment game), to distributions of equilibrium payoffs (at contract layer), to the evaluation of these distributions and market equilibria (at product layer), up to the point of fully fledged DSGE[12] models combining several markets (macroeconomic layer). It is an essential step in modelling to chose the suitable level of abstraction in order to make your point.

A visual representation of the structure of lending theory is provided by figure 2.1 on p. 16. Let us call it the 'map of lending theory'. In analogy to other subjects, it can be seen as a kind of

- lending theory analogue to component diagrams (similar to UML[13] diagrams in software engineering),
- or as an anatomy of the 'organs' of lending theory (like anatomic drawings in medical science),
- or as an exploded-view drawing of the 'lending machinery' (as in engineering science).

I will talk the reader through the details of the new structure step by step.

As soon as the new structure is established, we can work with it. The subsequently provided literature review in chapter 3 proceeds along the structure developed here. Likewise the whole argumentation in the reminder of this thesis is based on the mindset of layers of abstractions established here.

11 For example Nisan and Schocken confirm that "the articulation, use, and implementation of such abstractions are the bread and butter of our profession: Every hardware and software developer is routinely defining abstractions (also called "interfaces") ... . The abstractions are often built layer upon layer, resulting in higher and higher levels of capabilities." [NS05, p. 3].

12 Dynamic Stochastic General Equilibrium

13 Unified Modeling Language

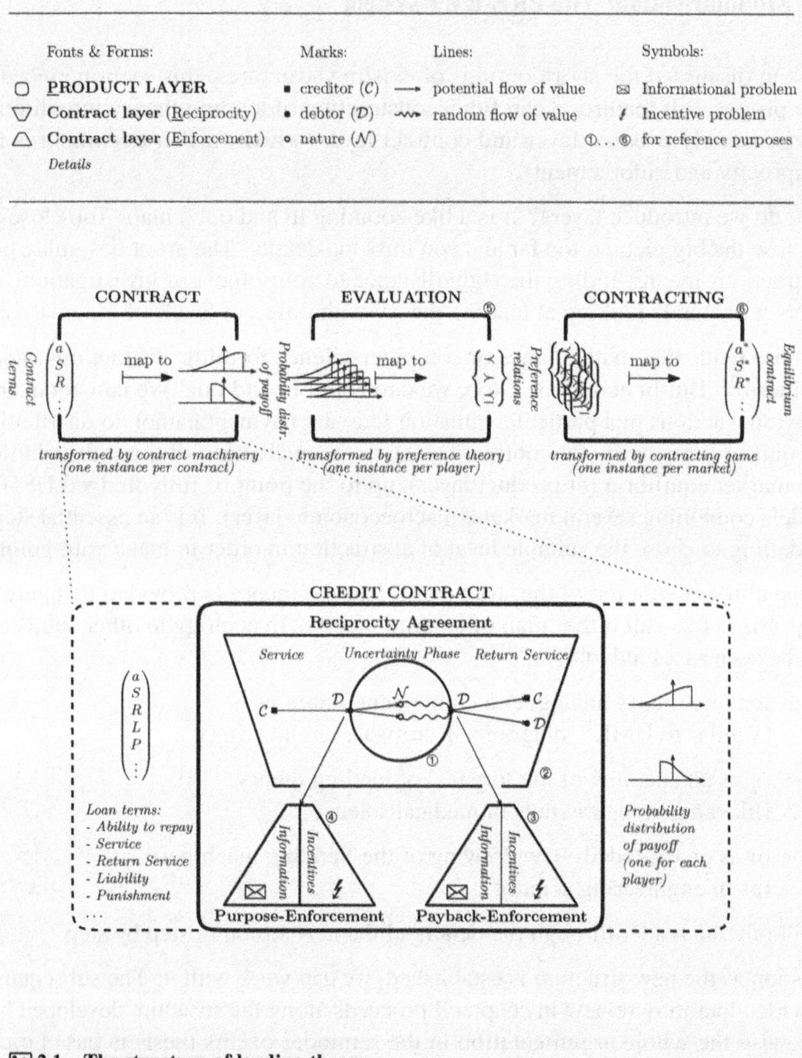

2.1 The structure of lending theory
A major achievement in this dissertation is the careful disentanglement of the interplay between contract features and market features. Thorough structuring is an important effort in order to equip practitioners with a clear picture of the many dependencies between different lending aspects and it is a necessary prerequisite for the implementation of a modular research strategy. Details are elaborated in the text.

2.3.1 Product-Layer

First of all, we separate topics that are not credit contract specific from the rest of the theory. On the product layer we work at a level of abstraction where we pretty much forget what product it is all about. Whether it is credit contracts, insurance contracts, second-hand cars or any other contract does not matter at this layer. Thereby we can talk about, discuss and research these things in a universal manner. Also the results aquired at this layer will be quite general and thereby applicable to a wide variety of different situations.

Contracting (⑥)[14] for example is a problem that appears in any market, just as informational problems like hidden characteristics are present in other markets, too – e.g. Akerlof's market for lemons [Ake70]. Many papers about adverse selection are just adaptions of the same mechanisms for different markets (e.g. credit markets, insurance markets, labour markets, lemons markets, ...).

Likewise, the variety in lending theory coming from evaluating the same credit contract (basically a probability distribution of payoff) with different utility functions can be treated separately. A player's *evaluation* (⑤) of a cash flow promise depends e.g. on his risk-attitude and reservation level of utility. Whether the probability distribution of payoff comes from an insurance contract, a credit contract or a lottery ticket is irrelevant at this layer.

The purpose of the *product layer* of lending theory is to separate *what* the contract does from *how* it does it. From this viewpoint a credit contract is just a product on a market – just a probability distribution of payoffs. How exactly the credit contract works and all the entrails remain a black box. Figure 2.2 illustrates what a credit contract looks like as seen from product layer.

Here, the probability distribution of payoff is not only the interface to the evaluation module, but also the dividing line between credit contract specific and unspecific topics; whether the underlying contract leading to a certain distribution of payoffs is a credit contract or e.g. an insurance contract is irrelevant for the evaluation and contracting module.

The structure is similar to other markets. In the market for lemons, a car with certain properties like type and kilometre reading leads to e.g. a probability distribution modelling its remaining service life. Lots of lending theory is actually theory about the functioning of markets in general – where credit contracts could be replaced by any other product with similar properties.

14 Circled numbers refer to figure 2.1 on page 16.

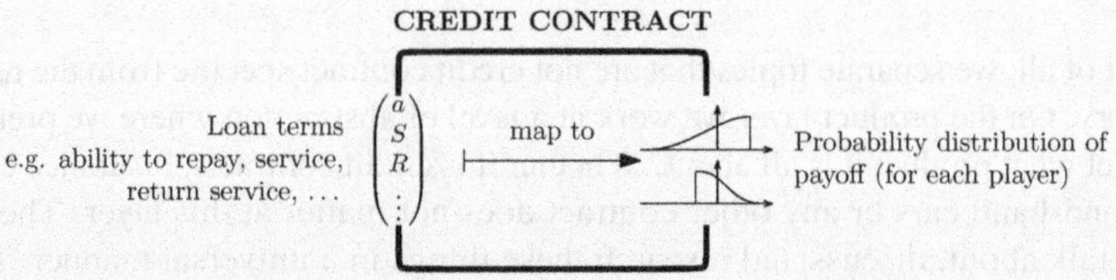

2.2 Credit contracts seen from product layer
Seen from product layer the credit contract is condensed to the information on how the loan terms manipulate the probability distributions of payoff for each contract party. How exactly the contract performs this manipulation remains a black box at this layer.

Q 2.1 Parametrising credit contracts
One could object that the whole structure of a credit contract can be subject to negotiations and just looking at parameters like interest rate R and collateral C does not meet the complexity of contracting. But notice that the structure of a contract is enclosed within these parameters. E.g. if we want to model the choice between collateral and no collateral we would build a model with a parameter C and set $C = 0$ to obtain a contract without collateral; likewise for joint liability structures and other aspects. The parameters reflect the structure of the contract. Or put differently: The variability in the structure of a credit contract ideally should be parametrised before analysing its effects.

2.3.2 Contract-Layer

At *contract layer* we gain a closer look at the machine room of credit contracts. A credit contract consists of its heart – the deferred reciprocity agreement (②) – plus additional parts (③,④) to enforce the deferred reciprocity agreement.

Examples for additional contract constituents are a contractually fixed utilisation of the money lent plus various monitoring schemes (see section 3.4), or regular reports and compulsory group meetings, or payback-enforcement means (see section 3.5) like auditing rules and penalty functions. Quite some variety in lending theory arises from replacing enforcement mechanisms. There can be several different means to enforce the same reciprocity agreement (see figure 13.2).

In a perfect world, without informational asymmetries and moral hazard, enforcement aspects may not be part of the credit contract at all. In contrast, the reciprocity agreement is a mandatory part of *every* credit contract by definition (see p. 3).

2.3.3 The layer's hierarchy, objectives and interfaces

When introducing the layers of lending theory (product-layer, reciprocity-layer, enforcement-layer), the storyline has been to – step by step – zoom in. First of all we look at the credit market and see a product (product layer). Then we open the box and find a credit contract, i.e. a deferred reciprocity agreement (contract layer: reciprocity). Further down into the fine-print we encounter additional contract constituents that exist to enforce the deferred reciprocity agreement (contract layer: enforcement).

Each of these layers has its dedicated function. Bare reciprocity agreements are only possible amongst intimates that share a bond of trust (see figure 1.3 on p. 8). The contract layer combines the contract's reciprocity agreement with enforcement means. It is the vehicle that renders deferred reciprocity agreements practicable for people who cannot rely on mutual trust. Eventually, the product level encapsulates the contract, makes it marketable and thus enables credit connections even between afore unknown people. These layers successively increase the applicability of deferred reciprocity agreements.

Let's conclude this section with an examination of the quality of demarcation of these layers. To do this we have to check if the interfaces between these layers are reasonably simple (see p. 12). But we have already seen that the product layer only requires the information on how loan terms are turned into probability distributions of payoff (p. 18). And concerning the reciprocity agreement, it only requires the information whether an enforcement module can enforce a certain action of a borrower at a certain decision point and how high the dead-weight costs of this enforcement game are. Thus the necessary interfaces between these layers are reasonably simple.

2.4 Modularisation: Implementation blackboxing

Blackboxing, i.e. encapsulating structural details in a black box is a very useful technique. In lending theory there are quite some opportunities to employ it. As we will see in the course of the chapter, often different assumptions (*implementations*) can have the same effect (*impact*) on the lending game. Therefore it is reasonable to abstract from these concrete implementations and focus on the induced impact (*blackboxing*). The consequent utilization of this principle can be regarded as a kind of argumentative hygiene. Again, we clearly separate the 'what' from the 'how'. Let's look at three lending theory specific examples.

2.4.1 Example: Ability to repay

Papers often start out with an investment project as a reason for borrowing:

2.2 The ability to repay in the literature

[Bes85, p. 851]	"Each entrepreneur has the opportunity to undertake a project which requires a fixed amount of investment *I*."
[BC95, p. 3]	"A borrower has a project that requires one unit of capital."
[Coc00, p. 193]	"Suppose that an entrepreneur owns an investment project with a two-points return distribution."
[DQFG16, p. 219]	"They [the borrowers] have access to a stochastic production technology that requires 1 unit of capital per period ..."

By doing so they confine themselves to a very specific lending situation – an investment loan. Borrowers, however, take loans to buy a house, to finance their education, or to use the proceeds for repayment of past debts – and according to Guérin et al. the portion is not at all irrelevant: "Depending on the zone and population, this [the repayment of past debts] represents between a tenth and a quarter of microcredit use" [GMRS11, p. 116].

Many reasons for borrowing exist. Certainly we ought not build a new model for every variation. So what is the actually relevant information of these assumptions? It is the answer to the question of whether the borrower will be able to repay! Although there are many possible situations, we have a single impact: A probability distribution representing the ability to repay. Potential implementations: consumer credits, investment credits, follow-up credits, different levels of wealth, etc. We separate the *cause* for a certain ability to repay from its *effect* on the rest of the model.

As a result we can harvest the benefits of specialization. Some specialists can focus on the challenging task of modelling probability distributions for the ability to repay given different usages of the loan. Whilst other researchers can focus on how the lending contract is affected for different abilities to repay.

2.4.2 *Example: Punishment*

Definition. Punishing means to deliberately decrease the utility of someone else, because they have done something wrong, e.g. broken an agreement.[15]

The definition of punishment could be the one state above. Thus, being able to punish someone means being able to modulate someone's utility. In the literature papers regularly bother with punishment implementations.

2.3 Punishment implementations in the literature

[Dia84, p. 396]	"We will see that these penalties are best interpreted as bankruptcy penalties. Some examples include a manager's time spent in bankruptcy proceedings, costly "explaining" of poor results, search costs of a fired manager, and (loosely) the manager's loss of "reputation" in bankruptcy. Physical punishment is a less realistic example."
[BC95, p. 4]	"The bank's penalties can be thought of as having two components. The first component is a monetary loss due to seizure of income or assets. … The second component is a non-pecuniary cost resulting from being 'hassled' by the bank, from loss of reputation, and so forth."
[Arm99, p. 82]	"We shall denote by V the value that an individual borrower derives from being able to re-access future refinancing. V is lost if the borrower does not meet her debt repayment obligation."
[RS04, p. 217]	"We assume that the bank can credibly threaten such punishment, even though it involves a deadweight loss. In practice, the Grameen Bank does impose non-pecuniary punishments, such as delaying or denying future loans … "

What does it mean to distinguish between implementation and impact here? It means to regard the effect of punishment, i.e. the reduction of the borrower's utility, separately from the concrete means by which the punishment is brought about, e.g. penalties, social sanctions, denial of future loans, etc.

15 The definition is loosely based on the OALD entry "punish (Verb)", see `http://www.oxfordlearnersdictionaries.com/definition/english/punish`.

2.4.3 *Example: Reservation level of utility*

Papers frequently determine the interest rate by exogenously fixing the amount of expected repayment. Possible explanations are:

- opportunity costs of capital (e.g. outside options at financial markets)
- costs of refinancing loans (e.g. interest rate for deposits)
- internal profit goals (e.g. set by stakeholders)

The impact of these implementations is always the same: The bank won't engage in a lending contract that yields an expected utility below the bank's reservation level of utility. In most cases it is enough to work with this impact, leaving bothering details about its implementation in a black box.

2.5 Modularisation: Enforcement modules

In this section we will have a special look at enforcement modules as they are a particularly important part of credit contract theory. What makes them special and interesting is a fundamental dichotomy. But first things first.

2.5.1 *Uncertainty phase decomposition and moral hazard*

When zooming in on the uncertainty phase of a credit contract it becomes obvious that it actually consists of different types of uncertainties (figure 2.3). Between service and return service we typically find two borrower decision points. And there is some inherent fortuitousness that models whether the borrower earns enough such that he can repay – which can be taken as a decision point of 'nature'. Enforcement is an issue that originates at such borrower decision points. Typically a credit contract includes two potential borrower decisions: utilisation of the loan and repayment. But there can be more than the two standard ones! A better wording than the unspecific and fixpoint-dependent terms 'ex ante' and 'ex post' thus refers to the exact decision point in the lending game where the enforcement problem is born. Examples are:

- purpose-enforcement (the borrower shall use the loan as stipulated)
- payback-enforcement (the borrower shall repay the loan)
- liability-enforcement (the borrower shall bail out his peer)
- monitoring-enforcement (the borrower shall monitor his peer)

By this convention we gain a precise way of speaking about enforcement problems in lending theory which is independent of the timing of our models.[16]

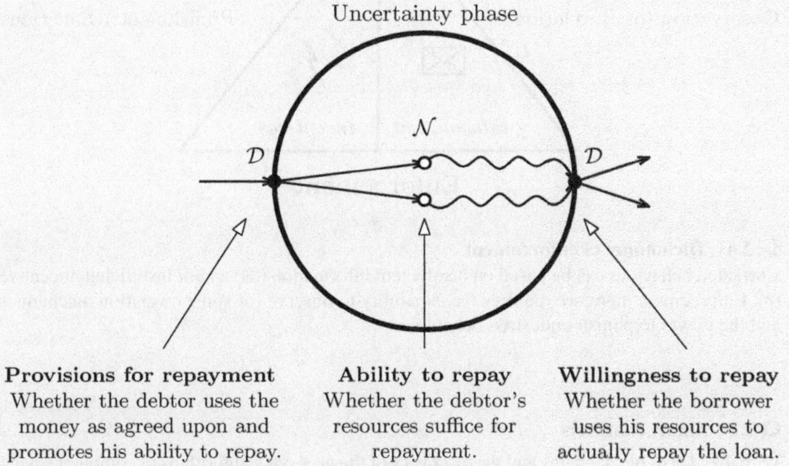

2.3 Uncertainty phase decomposition
A closer look at the uncertainty phase of a credit contract reveals that the uncertainty is composed of multiple issues.

2.5.2 *The two halves of enforcement modules*

To analyse how enforcement is achieved, we need enforcement modules (③, ④ in figure 2.1). It is a crucial insight that effective enforcement always requires two different capabilities: The undesired action has to be identified (informational aspect) and you need to have sanctions at hand that are strong enough to prevent it (incentive aspect). We cannot treat these two aspects separately. For example purpose-enforcement requires a mechanism to gain information and a sanctioning mechanism if the diversion of funds is revealed. Likewise, to prevent strategic default, the lender must, on the one hand, find out whether the borrower actually can repay and, on the other hand, threaten him with some kind of punishment. Conclusion being: The capability to sanction borrowers without the ability to discover moral hazard is as

16 The terms 'ex ante' and 'ex post', in contrast, reference the time line of a model. To give an example: Depending on the peer monitoring effort being decided before or after the realization of some random variable, the same issue may be an instance of ex ante or ex post moral hazard.

useless as the ability to recognise unwanted behaviour without the possibility to sanction it sufficiently. Figure 2.4 illustrates this dichotomy.[17]

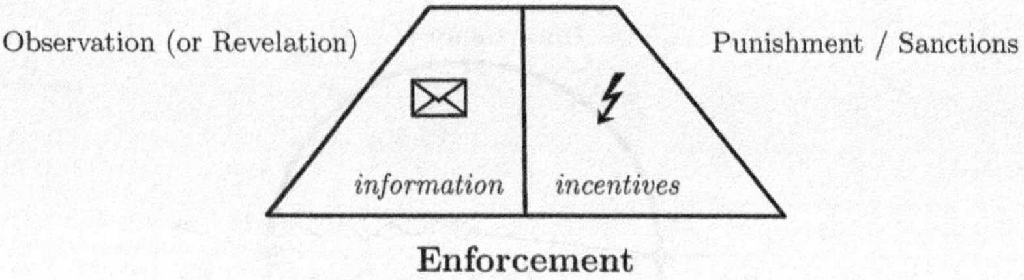

2.4 Dichotomy of enforcement
Unwanted behaviour can be based on insufficient information (⊠) and/or insufficient incentives (↯). Enforcement therefore requires the capability to observe (or some revelation mechanism) and the power to punish undesired behaviour.

Q 2.2 Smart contracts
With the idea of block-chains and the discovery of the necessary algorithms in computer science the concept of smart contracts came up. A pioneer in the field is the computer scientist Nick Szabo who defines: "A smart contract is a computerized transaction protocol that executes the terms of a contract."[a]
So what does a smart contract do from a theoretical perspective? Actually, smart contracts could eliminate the repayment decision node (see figure 2.3) right away. In the fashion of a vending machine the contract is executed automatically – in our case a deferred reciprocity agreement. Quite enthusiastically Szabo writes: "This gives us new ways to formalize and secure digital relationships which are far more functional than their inanimate paper-based ancestors." [Sza97] So in the future, the staircase in figure 1.3 may experience a further extension by a form of deferred reciprocity between unrelated people, realised through credits based on self-executing smart contracts.
However, up to now, many issues remain open. Most importantly the question arises which are the conditions we can contract on with this new technique? Obviously, some things can be digitised easier (e.g. bank account balances) than others (e.g. effort levels). The algorithm would need to have full information on the relevant states of the world; also manipulation of a smart contract must be prohibitively costly. Furthermore, the necessary infrastructure and know-how to write smart contracts may not be available in every part of the world. However, in markets that are sufficiently digitised, maybe, strategic default soon belongs to history.

a `http://www.fon.hum.uva.nl/rob/Courses/InformationInSpeech/CDROM/Literature/LOTwinterschool2006/szabo.best.vwh.net/smart.contracts.html` 2018-07-11, 17:00.

17 Note that the term 'enforcement' is used as an umbrella term to refer to informational *and* incentive issues here. Both must intertwine to achieve enforcement. The screening-monitoring-auditing-enforcement-structure of Fischer and Ghatak (section 2.2), reserves the term enforcement exclusively for incentives aspects.

3 Specifications: The modules and their treatment in the literature

This chapter provides a literature review. It is, however, not a standard review that takes individual publications and repeats their results. In contrast, its innovation is to proceed in a cross-sectional fashion. It restructures the content along the modules of lending theory identified in the previous chapter (see 2.1). The emphasis lies on carving out the fundamental working principles of each module.

After a preliminary discussion of lending tools in section 3.1, the sections are:

3.2	**Deferred reciprocity agreement**	②
3.3	**Ability to repay**	①
3.4	**Purpose-enforcement**	④
3.5	**Payback-enforcement**	③
3.6	**Evaluation**	⑤
3.7	**Contracting**	⑥

Each section, on the one hand, presents the fundamental mechanics of every module and, on the other hand, sketches how researchers have modelled it in the associated theoretical literature. Eventually, the whole review will be completed by a look at the vicinity of credit contract theory in section 3.8.

3.1 Lending Tools

3.1.1 Information tools

In this subsection we briefly focus on the informational aspects of lending theory. There are three asymmetric information issues: *hidden characteristics*, *hidden actions* and *hidden states*. We can of course regard them as three comparable topics, as figure 3.1 suggests. But mentioning the related issues of screening, monitoring, auditing in a row suggests a structural similarity that in fact is not existent. The anatomy of lending theory reveals that these issues occur at different layers (see figure 2.1 on p. 16). Whilst hidden characteristics are a problem of contracting games at the product layer of lending theory, hidden actions and hidden states occur at the

C. Prem, *The Theory of Credit Contracts*,
https://doi.org/10.1007/978-3-658-29362-8_3

contract layer. And whilst hidden actions and hidden states are two prerequisites for the occurrence of moral hazard, hidden characteristics do not induce moral hazard.

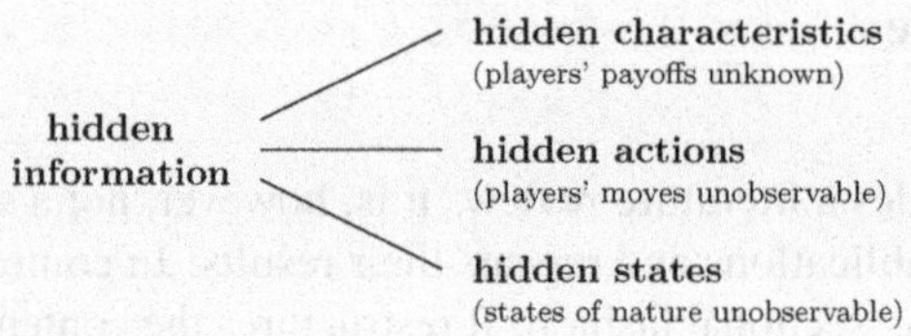

3.1 Three types of hidden information
Conceptually, we can distinguish three different types of hidden information.

3.1 Incomplete information, imperfect information, Bayesian games

How does the wording used in figure 3.1 match with the game theoretic language of incomplete information, imperfect information and Bayesian games?

With *incomplete information* game theorists refer to games where players do not have all relevant information about each other (e.g. each other's wealth or risk aversion). In particular the payoffs in various outcomes of the game are not commonly known [MCWG95, p. 253]. A more succinct wording: **hidden characteristics.**

The term *imperfect information* describes games where some players cannot observe other players' moves, i.e. some information sets are not singletons [MCWG95, p. 270]. A pithy synonym: **hidden actions.**

Bayesian games are games with a random player, called '*nature player*', whose random decisions are not publicly observable. The hidden actions of nature are referred to as **hidden states**. There is in fact a disparity between hidden actions and hidden states: The random player does not act strategically, the probability distribution is given exogenously. As Mas-Colell et al. mention, sometimes "the situation of incomplete information is reinterpreted as a game of imperfect information" [MCWG95, p. 254] by using Bayesian games. Put differently the quotation says hidden states can be employed to model beliefs about hidden characteristics.

Furthermore we should thoroughly distinguish between causes and problems and solutions here. E.g. the literature frequently speaks of the 'problem' of screening (e.g. see [FG11, p. 63]), but screening, of course, is a potential *solution* – amongst others – to tackle hidden characteristics. Also using the term monitoring "in a broad sense" [FR08, p. 30] to refer to screening, the prevention of ex ante moral hazard, punishing and auditing is not at all helpful.

To clarify the technical terms, table 3.2 provides an overview and assigns each term its appropriate place. Its last row collects the lending tools that deal with the respective asymmetric information issues of lending theory.

Informational Asymmetry	hidden characteristics	hidden actions	hidden states
Affected Layer	product layer	contract layer	contract layer
Associated Problems	adverse selection, credit rationing	diversion of funds, low effort	strategic default
Possible Countermeasures	screening, signalling, self-selection	lender monitoring, delegated monitoring, peer monitoring	auditing, cross-reporting

Remark: Auditing is also termed costly state verification.

3.2 Asymmetric information: Types, Problems, Tools
This overview concerning informational asymmetries in lending theory thoroughly differentiates between causes, problems and tools.

3.1.2 Incentive tools

The previous subsection talked about the information part of enforcement (see p. 24). Now we look at the incentive part of enforcement. How can creditors set incentives? A list of tools is provided below.

Pecuniary fines for violating contract terms e.g. repaying too late
(requires borrower wealth that exceeds the stipulated return service)

Legal consequences like lawsuits and prison
(requires a working judicial system)

Complete financial exclusion e.g. via reports to a consumer reporting agency[18]
(only bites if further financial services are needed)

Reduced value of future loans e.g. by making future loans more expensive, delaying future loans or restricting their size
(builds on the absence of attractive alternative sources for credit)

Restricted access to saving
(bites if the bank offers a particularly attractive way to save)

Loss of reputation in public, e.g. as a consequence of public repayment
(uses social networks and social pressure as incentive)

18 Notice the two aspects of consumer reporting agencies:
1) They can be seen as screening tool. 2) They can be used as a punishment device.

Even more variations arise if you combine different means. Also, the sanctions need not be connected to finance. In China's Social Credit System the consequence of not repaying is the denial of train and flight tickets.[19]

Often collateral is mentioned as a tool to enforce repayment. However, we should regard it as what it is, namely a part of the ability to repay (see p. 42) and not an incentive tool. It enables a transfer flow (p. 103) from borrower to lender which may need to be enforced itself. Nevertheless, in the case of asymmetric valuation, i.e. if the collateral is more valuable to the borrower than to the lender, we must conceptually distinguish between the value actually transferred to the lender and the additional value lost by the borrower. Only the value lost can be seen as a pure punishment and hence as an incentive tool.

Under implementation blackboxing (see section 2.4) all the details of these incentive-setting-vehicles disappear in a black box. The achieved intensity of the incentive is the simple interface connecting it to the rest of the lending game. Complexity control is important here, as the implementation of incentive tools can become a rather complex sub-problem! The reason being that the incentive effect of some of the tools mentioned above is relatively hard to calculate. For instance social sanctions, like the loss of reputation, will hit different personalities with differing severity and may exhibit tremendous interpersonal variability. But also the intensity of seemingly more concrete tools like the denial of future loans is highly dependent on a borrower's need for future financing and the availability of alternative sources of funding. Bond and Rai have already pointed at these issues in 2002: "Successfully imposing social sanctions requires the successful navigation of a delegation problem, while credit denial lacks a market value … ." [BR02, Abstract] In addition, it is not only the severity of these threats when they strike but also their ex ante credibility that is relevant for the achieved incentive effect. E.g., for a bank, denying financial services also means loosing potential customers. As D. Gromb mentions, the bank's credibility depends on the borrower's profitability: "If the future relationship between the lender and the borrower is profitable, the lender has little incentive to put it to an end ex-post, i.e. after default. Hence, ex-ante the threat of denying the borrower new capital is weakened." [Gro94, p. 2] So quite obviously some of the incentive tools above require an extra block of theory.

19 German original: "Wer seinen Kredit nicht zurückzahlt … der darf in China in vielen Fällen bereits nicht mehr mit dem Schnellzug oder mit dem Flugzeug reisen." From the article "Chinas Sozialkredit-System – Auf dem Weg in die IT-Diktatur", Deutschlandfunk Kultur. Link: `http://www.deutschlandfunkkultur.de/chinas-sozialkredit-system-auf-dem-weg-in-die-it-diktatur.979.de.html?dram:article_id=395126` Accessed 2018-Jun-22, 16:50.

Social sanctions, for example, need a theoretic model of social structures and the value of social ties. Postelnicu, Hermes and Szafarz present an interesting article where they tackle the conceptualisation of social capital with a comprehensive analysis of the different types of social ties in a social network (see [PHS14]). Baland, Somanathan and Wahhaj try to determine the intensity of non-pecuniary social sanctions by calculating the loss that comes from exclusion from communal activity (see [BSW14]). De Quidt, Fetzer, and Ghatak conceptualise social capital as the lifetime payoffs in a repeated coordination game which stands for a simple model of friendship (see [DQFG16, Section 2.4]). Concerning financial exclusion, Armendáriz takes a stab in calculating the value of a non-refinancing penalty (see [Arm99, Appendix A]). Bond and Krishnamurthy comprehensively discuss and compare various forms and levels of financial exclusion (see [BK04]). And Gromb examines how limited credibility reduces the effectiveness of threats. [Gro94, p. 2]

3.1.3 Microlending tools

After the compendious overviews in sections 3.1.1 and 3.1.2 let's look at information and incentive tools used in Microfinance. It is a preparation for the following sections, as they will always throw a side glance at microfinance.

First of all: What is meant by the terms microfinance, microcredit, group lending, and joint liability? A definition of microfinance can be found in [Imb05]: ""Microfinance" is the provision of loans, savings, insurance, payments and other basic financial services to low-income populations." [Imb05, p. 67] Whilst in the beginning microfinance used to have a focus on lending, nowadays the term microfinance indeed refers to a whole bunch of financial services like microcredit, microsaving or microinsurance. Amongst these, microcredit is the service relevant in this monograph. (For details on microsaving and micorinsurance see [AM10, Chapter 6] or specifically on microinsurance [Chu11].)

But careful, the term *micro*credit is somewhat misleading. Though loan size can be small, the labelling is usually used to refer to a variety of innovations that are not necessarily linked to loan size. Like pharmaceuticals with various modes of action, these microcredit tools target at different aspects of the anatomy of lending presented in figure 2.1 on p. 16. In the list below the targets are specified.[20]

20 Also notices the lists of lending tools produced by other authors, e.g.
[AM10, Chapter 5, "Beyond Group Lending"],
[FR08, Section 4.3, "Incentives to Repay"],
[Led98, p. 137-138, "Collateral Substitutes" & "Alternative Forms of Collateral"].

Lending combined with business training helps to ameliorate business decisions, hence improves the borrowers' ability to repay (target: ①). Therefore some microcredit programs include additional business training [KV11].

Progressive sequential lending is a special form of *sequential lending* (without progression). It makes future credits conditional on the repayment of the current one and increases loan size over time. The progression boosts the incentive effect. The debtor builds up valuable reputation whose potential loss can[21] serve as an incentive to repay (target: ↯ of ③).[22]

Frequent repayment is a component of most microfinance contracts. Even weekly collection of repayment instalments is widely used (see [FP08]). The idea is to split and thus defuse the uncertainty phase (target: ①) by fragmenting the return service into several parts.[23] For Rutherford "the opportunity to save or repay in frequent bite-sized amounts is the main driver of microfinance's success – far more important than other better-known aspects such as group-formation, joint liability, and the focus on investments in microenterprises." [Rut11, p. 517]

Public repayment can improve repayment rates by "the ability to generate stigma" [AM10, p. 157]. Seen from that "stigma"-perspective, it simply serves as an incentive tool (target: ↯ of ③). Another perspective regards public repayment as a precondition for implicit joint liability. Making repayments publicly observable enables the villagers to contract on their peers' repayments ([RS13, p. 260]) and thus enables side-contracting amongst borrowers. In this context public repayment is interpreted as an information tool.

Targeting women refers to giving credit predominantly to female borrowers, as empiric evidence indicates higher repayment rates compared to men (see [AM10, p. 158]). The underlying reasons remain vague. Maybe women are more trustworthy concerning ex ante moral hazard (target: ④), perhaps women make better business decisions (target: ①, ability to repay). A side glance into anthropology, however, reveals that it could be an exploitation of women's humble social position making them more vulnerable to pressure (target: ↯ of ③ and ④) exerted by bank workers and peers (see [Rah99]).

21 If the debtor has no further needs for financing, this incentive does not bite.

22 Sequential lending, and its amplification progressive lending, are also subsumed under the term *dynamic incentives* [AM10, p. 143]. Others term the idea *contingent renewal* [Cho07].

23 For a discussion of repayment frequency, its benefits and drawbacks like increased transaction costs, see [AM10, Section 5.3]. In addition to the arguments considered by [AM10], high-frequency repayment can be particularly useful in dealing with present-biased borrowers [FG10]. However, also consult [AB13] for a flip side to the argument that microfinance helps present-biased borrowers.

Of course, some of these 'innovations' promoted under the label microcredit are not at all microcredit specific. Dynamic incentives, for instance, are prevalent outside the world of microcredit, too, when borrowers are rated by credit reference agencies: A bad rating usually leads to denial of future loans or at least worsens credit conditions for the borrowers concerned. Progressive lending exhibits similarities with so-called *relationship banking* [FR08, p. 99]. And frequent repayment is just an extreme form of paying by instalments.

Another set of tools ranging under the general term microcredit is *group lending*. Unfortunately, the label is blurred. It is especially important not to use group lending as a synonym for joint liability. Group lending can refer to a number of different mechanisms that create dependencies between individual borrowers. Some are listed below.

Sequential group lending is another form of sequential lending. Instead of making future credits of just one borrower conditional on his repayment, the financing of a whole number of borrowers is put at stake. The other borrowers of the group only receive credit if the current debtor repays, which increases the pressure to repay by a peer pressure component (see [Cho05, Cho07, Ani03]). Like progressive sequential lending, this is a way to boost the incentive effect (target: ↯ of ③) of plain sequential lending.

Cross-reporting is based on an assumed informational advantage borrowers have compared to the bank (target: ✉ of ③). The process of repayment[24] is enriched with a message game where the borrowers reveal information about their peers' actual ability to repay (see [RS13, p. 259, footnote 11] and [RS04, p. 230]).

Joint liability is a form of group lending where several borrowers help each other in completing their return services. The primary effect of joint liability is a risk sharing effect (target: ①, ability to repay). We can distinguish different forms of joint liability. According to [DQFG16, p. 220], joint liability can be hard-coded into the credit contract by the bank (*explicit* joint liability) or can be planned privately with "repayment guarantee arrangements" between borrowers (*implicit* joint liability, also called *side-contracting*). Furthermore, the liability for peers can be full or partial, see [All16].

Since group lending in general makes borrowers dependent on each other, it induces a number of secondary effects. It "incites members to perform various functions, including screening of loan applicants; monitoring the individual borrower's efforts, fortunes, and shocks; and enforcing repayment of their peers' loans." [Zel98, p. 599]

24 The borrowers must make their repayment to the bank 'simultaneously', which is the game theoretic formulation for 'without knowing their peer's action beforehand'.

Clearly, Zellner refers to information and incentive issues (indirect targets: ⊠ of ⑥ and ④, ⚡ of ③). According to his statement group lending arrangements work like a transmission belt to transfer parts of the lenders' tasks to the group of borrowers.

But for the creditor, not everything comes easy with group lending. Enforcing a joint liability contract and preventing strategic default can be harder than enforcing an individual lending contract.[25] Also collusion against the bank is a prevalent issue. And seen from a welfare perspective the question is, whether borrowers can really accomplish the tasks mentioned above more efficiently than a professional bank.

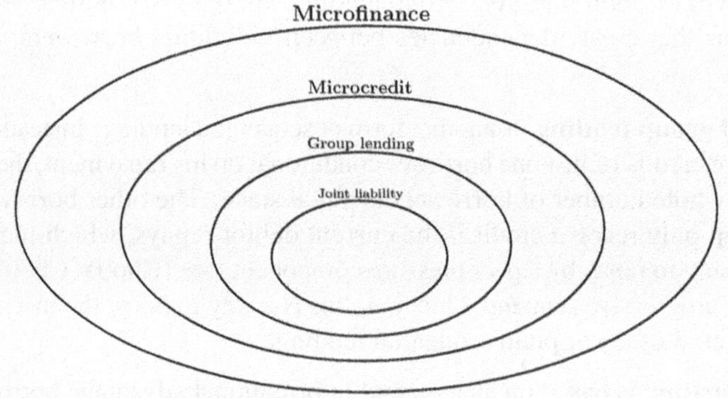

3.3 **Zooming into Microfinance**
Joint liability is a group lending tool in microcredit which in turn is one aspect of microfinance.

3.2 The reciprocity agreement

The structural overview in figure 2.1 on p. 16 presented a very stylised view on deferred reciprocity agreements. Only the directions of potential flows of value were specified. In this section we will have a closer look on these flows, their value and type.

25 Compare the two contracts B.1 and B.5, which exhibit the same punishment intensity. Obviously, enforcing a joint liability contract requires a higher punishment intensity, see B.2.

3.2.1 Formalization of deferred reciprocity agreements

The variety of deferred reciprocity agreements leads to a large number of different forms of credit contracts. Other authors also try to sort through this dazzling array (e.g. see figure 1.2). We establish order systematically, using the definition on page 4. For a deferred reciprocity agreement, the following loan terms are essential:

1) Service: Provision of credit and the value of the loan

2) Duration: The time lag between provision of credit and return service

3) Return service: Gross interest (i.e. principal plus interest) and collateral

Service

To fix the service part of a credit contract, two things must be defined. The first one is the provision of credit. Of course the credit can simply be provided in cash. But it can also be, for example, a trade credit and does not need to be cash. So credit contracts vary with respect to the provision of credit. The second thing is the size of the loan. There is a huge variety, ranging from large public or corporate debts, over 'normal' size personal loans to tiny micro credits. In the literature, the form of provision of credit usually remains disregarded and models just work with its value. If the size of the loan is an invariant, it is frequently just normalised to one. However, there is some theory that explicitly examines the role of credit volume (e.g. [Han01, Section 2.3], [Sch01, KP98]).

Duration

Since we deal with *deferred* reciprocity arrangements, the time lag between service and return service is of importance. The duration of the uncertainty period must be stipulated in a credit contract. In many models duration is not modelled explicitly. Most of the time we have discrete periods, the time lag's duration is disregarded and in the repayment period the borrower repays in one go. However, in practice many contracts exhibit repayment by instalments. There is not much theory on such fragmentation of the uncertainty phase. However, it may be beneficial for the lender in the presence of present-biased, quasi-hyperbolic borrowers (see [FG10]).

Return service

Finally, the promised return service must be specified. This aspect of a deferred reciprocity agreement has received considerably more attention in the literature than the size of loans and their duration. The desired return service – in the context of bank loans called *gross interest* – typically consists of returning the principal lent plus some form of interest. Obviously, the distinction between principal and interest is only necessary if the difference between service and return service is of explicit relevance. In this case the interest is often reported by an interest *rate per year*. The advantage: Declaring the interest as a rate makes loans of different size comparable. And by specifying the interest rate per time, loans of varying duration become comparable. In models where neither size nor duration is varied, this distinction and normalisation is needless.

What determines the interest rate of a bank loan? In the literature several components of interest rates have been carved out. The return service of a bank loan may include:

Refinancing costs: the interest a bank has to pay itself on deposits

Profit margin: the lender's profit

Administration expenses: contract specific transaction costs as well as general fixed costs the lender has to deal with

Risk premium: an additional charge to account for the risk of default (this can comprise individual risks of a certain borrower as well as systemic risks)

Inflation offset: an additional mark-on to compensate inflation (turns nominal interest rate into real rate)

Be aware of the difference between inflation offset and inflation risk; the later refers to deviations from expected inflation if inflation is modelled stochastically. In a nutshell: The risk premium compensates the *uncertainty* in the uncertainty phase, the inflation offset compensates the *duration* of the uncertainty phase.

Note that each return service component serves a particular role. The first component shows the interest rate's role as the price for capital which moderates supply and demand at credit markets. The second component reflects the lender's bargaining power. The fourth component highlights the fact that the interest rate is an indicator for the riskiness of a loan. All these components finally mix with administrative expenses by the lender, which become particularly relevant in microfinance where loans are small with transaction costs comparatively high. So interest rates can be high due to high refinancing costs, high risks, high inflation, costly administration expenses or generous profit margins.

In practice, of course, nature provides much more possible outcomes than perfect success of a borrower, so simply specifying a single value as a return service is not sufficient. Instead, a credit contract must differentiate between different states; in fact, the return service must be specified for *every* possible state (see [FR08, p. 127]). That is why we need repayment functions.

Definition (Repayment function). A *repayment function* is a real-valued function specifying the stipulated value of the return service in every possible state of the world.

Notice that the repayment function is the return service *as promised* in the credit contract – not necessarily the actual repayment of a borrower. Whether the state is observable by the lender and the stipulated repayment is enforceable is a matter of the payback-enforcement module.

The codomain of a repayment function is $\mathbb{R}$. But what does its domain look like? The promised repayment can be contingent on the borrower's available resources, but the return service can of course be made conditional on other parameters, too (e.g. inflation-indexed bonds, joint liability contracts). Under costly state verification, for example, repayment may be made contingent on the borrower's reports (see [FR08, p. 130]). The domain of a repayment function can therefore be of various different forms, e.g. a discrete set, an interval, or a multidimensional space of states of the world.

Obviously, the specification of the return service (i.e. the repayment function) adds a considerable amount to the variety of credit contracts. Figures 3.4 and 3.5 provide two examples for repayment functions from the literature (compare [Coc00, p. 195, x196] and [SW81, Figure 2b]). Figure 3.5 shows the repayment function of a so-called *standard debt contract* (SDC), with collateral. It is characterised by a maximum which is constant for high values of the ability to repay and a return service that hands over all available resources (e.g. cash and collateral) to the creditor for low values of the ability to repay. In the literature this definition of a SDC can be found in [FR08, p. 132], [GH85, p. 648] and [KV00, p. 125].[26] Its popularity comes from the fact that it is amongst the repayment functions that can be enforced most efficiently (see [FR08, p. 135], [GH85, Proposition 3], [BS94], and section 3.5.3 of this thesis).

26 Notice that in [KV00] the acronym SDC is read as *simple debt contract*.

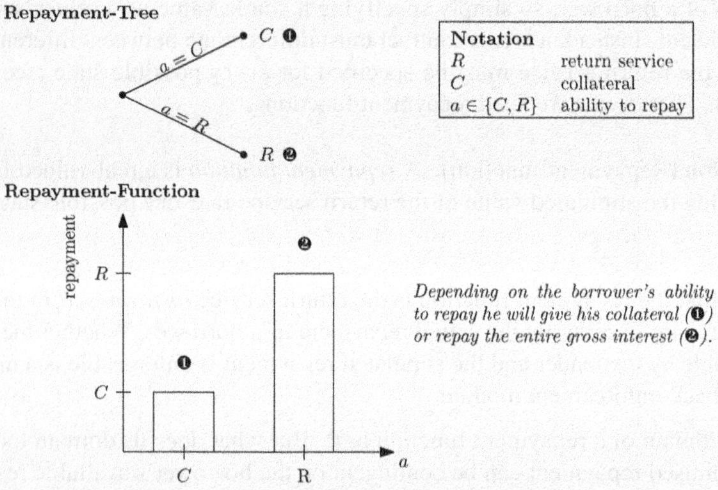

3.4 Repayment function for individual liability lending (discrete)
Given a world where the only two states are success and failure the repayment function of a reciprocity agreement must specify the return service in two cases.

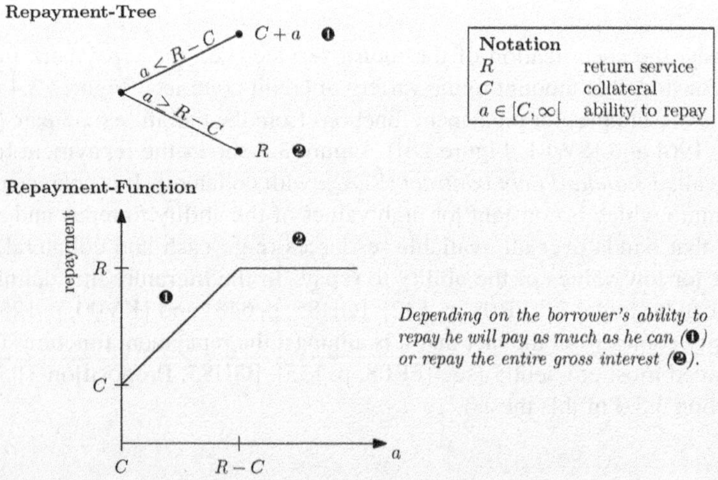

3.5 Repayment function for individual liability lending (continuous)
Given a world with infinitely many states of nature, the repayment function of a reciprocity agreements must specify the repayment in each of these cases.

3.2.2 *Special case: Joint liability contracts*

According to the literature joint liability can be explicit or implicit ([DQFG16, p. 217]). Figure 3.6 displays the flow-of-value-vectors in a joint liability contract. Notice that the triangle in the chart seems to be commutative, i.e. it seems to be irrelevant along which path the the values flow (e.g. $\mathcal{D}_1 \to \mathcal{D}_2 \to \mathcal{C} \equiv \mathcal{D}_1 \to \mathcal{C}$). However, if repayment is a decision node, then a borrower may choose strategic default although he is receives transfer payments from his peer. So unless transfer payments are automatically used for repayment implicit and explicit joint liability may not be equivalent anymore.

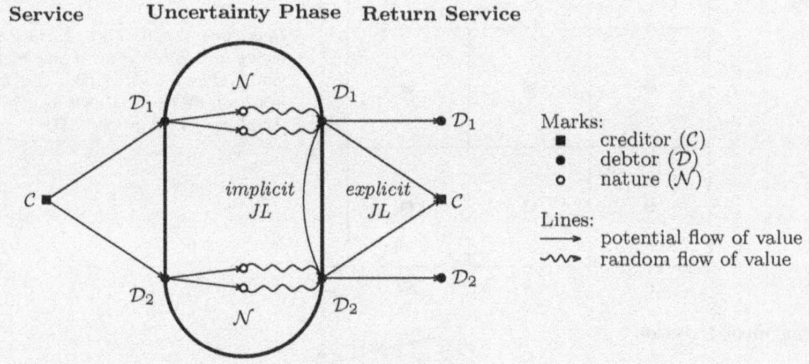

3.6 Reciprocity agreement with joint liability
The graphic illustrates the flow-of-value-vectors in a reciprocity agreement with two borrowers and joint liability (compare figure 2.1 on p. 16). Under implicit joint liability the borrowers transfer resources between each other. Under explicit joint liability the lender requests an increased return service if a bailout is necessary.

Since joint liability leads to a richer landscape of different situations – the stipulated repayment of a borrower not only depends on his own ability to repay, but also hinges on his peer's ability to repay – the repayment functions get more complicated, too. Figure 3.7 presents a possible repayment function for a borrower under joint liability lending.

And of course – although totally uncommon in the literature – the repayment functions of joint liability contracts can also be modelled continuously! Moreover, joint liability may be full or partial (see [BO10, All16]). And ultimately, joint liability may even be combined with collateral. Figure 3.8 demonstrates what the repayment function of such a credit contract could look like.

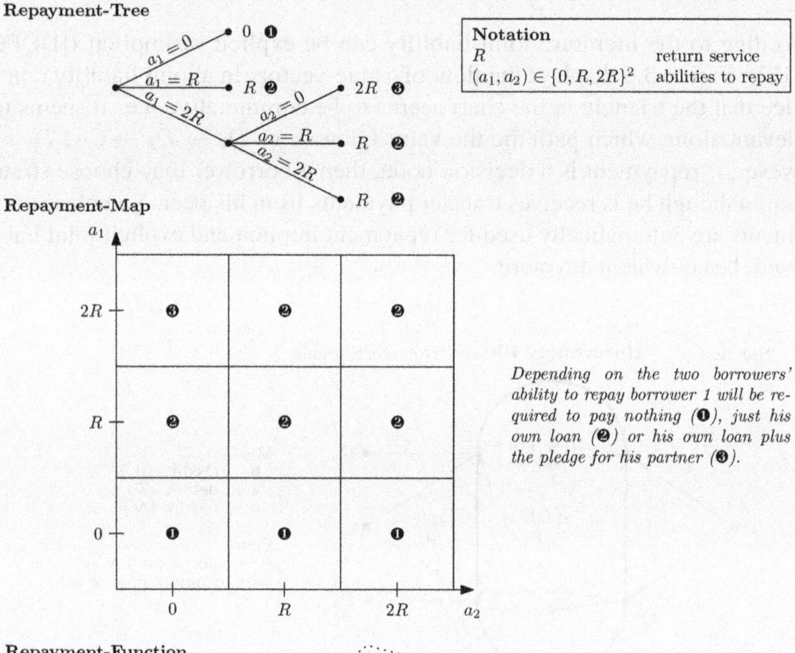

3.7 Repayment function for joint liability lending (discrete)
This figure shows – for a discrete world – the repayment that is expected of a borrower in a joint liability debt contract. The repayment function can be displayed as tree, map or function.

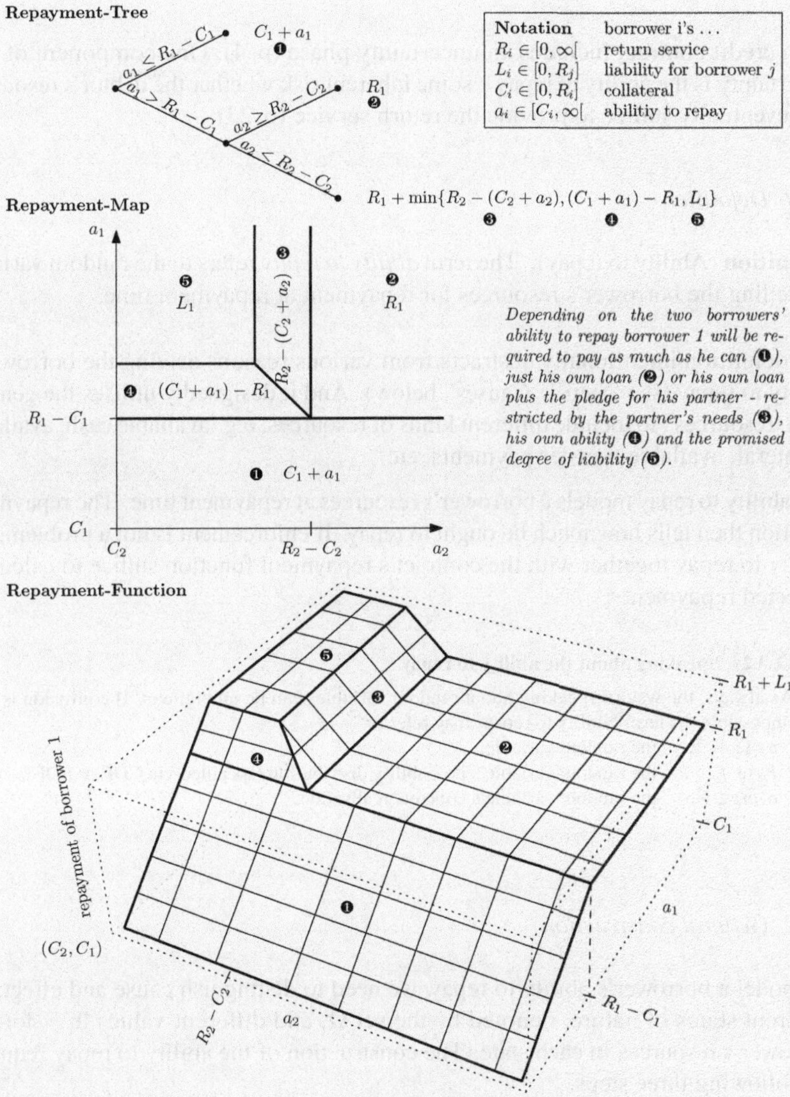

3.8 Repayment function for joint liability lending (continuous)
This figure shows what the repayment function for a borrower could look like in a continuous world. It features (partial) joint liability between borrowers, restricted to an amount of L_i, and it combines it with collateral C_i.

3.3 The ability to repay

Each credit contract includes an uncertainty phase (p. 4). One component of this uncertainty is the ability to repay – some inherent risk whether the debtor's resources will eventually suffice to provide the return service (p. 23).

3.3.1 Definition

Definition (Ability to repay). The term *ability to repay* refers to the random variable modelling the borrower's resources for repayment at repayment time.

The definition intentionally abstracts from various reasons driving the borrower's ability to repay (see 'Step 1: Causes', below). And it designedly utilises the general term 'resources', to include different kinds of resources, e.g. available cash, available collateral, available transfer payments, etc.

The ability to repay models a borrower's resources at repayment time. The repayment function then tells how much he ought to repay. If enforcement is not a problem, the ability to repay together with the contract's repayment function suffice to calculate expected repayment.[27]

Q 3.2 Speaking about the ability to repay

As always, the way of speaking about random variables can be ambiguous. If confusion is impossible the term 'ability to repay' may refer to

$a\colon \Omega \to \mathbb{R}$	the random variable,
F_a or f_a	the random variable's probability distribution (specified via CDF or PDF),
$a(\omega) \in \mathbb{R}$	the random variable's concrete realisation.

3.3.2 General construction

To model a borrower's ability to repay, we need to distinguish cause and effect, i.e. different states of nature, denoted by the set Ω, and different values in $\mathbb{R}$ for the borrower's resources in each state. The construction of the ability to repay requires the following three steps.

27 If enforcement is an issue, we need the equilibria of the enforcement game instead of the repayment function. See section 3.5.

Step 1: Modelling causes

The states of the world, captured by Ω, can comprise many different things like the outcomes of an investment project (investment credit), or a borrower's salaries (credit granted for consumption), or any other influences on a borrower's available resources at repayment time.

Step 2: Modelling resources

$$a\colon \Omega \to \mathbb{R}$$

The map a maps each state of nature to its associated amount of available resources. It can only take non-negative values since 'negative resources for repayment' semantically make no sense. Different states of nature may lead to the same amount of resources, so a is not necessarily an injective map and thus it is not simply a bijection of Ω onto a subset of $\mathbb{R}$. Otherwise the distinction between states of nature and available resources were rather meaningless.

Step 3: Modelling uncertainty

The sample space Ω, enriched with a σ-algebra Σ and a probability measure P, becomes a probability space

$$(\Omega, \Sigma, P)$$

modelling some real-world process associated with the different states of nature. The demand on a being a measurable map, makes it a random variable

$$a\colon (\Omega, \Sigma, P) \to (\mathbb{R}, \mathcal{B}(\mathbb{R}))$$

along which we can pull back the probability measure from (Ω, Σ, P) to $(\mathbb{R}, \mathcal{B}(\mathbb{R}))$.[28] We end up with a probability distribution A on $\mathbb{R}$ modelling the probability for each amount of resources. Since a is a real-valued random variable, its probability distribution A can be conveniently specified by a probability density function (PDF[29]) or in the form of a cumulative distribution function (CDF).[30]

28 $\mathcal{B}(\mathbb{R})$ being the Borel algebra on $\mathbb{R}$. If a model involves the calculation of expected repayment, a should also be P-integrable to ensure the existence of its expected value.
29 Also called a probability mass function if the random variable is a discrete one.
30 Notice that every real valued random variable has a CDF. But only probability measures with a CDF that is absolutely continuous possess a PDF.

Of course the statistically adequate construction of the probability space (Ω, Σ, P) is a challenge on its own. What is of relevance for the lending game is the probability distribution A on $\mathbb{R}$. Again, blackboxing leaves the details of constructing a realistic (Ω, Σ, P) within the box and just works with the interface.

3.3.3 The ability to repay in the literature

In the lending literature with individual liability we find two versions of modelling available resources. One possibility is the discrete modelling of A with a Bernoulli distribution. An example is the "simplified S-W type of model" in [Coc00, p. 195]. Most often, however, available resources are modelled continuously and specified by a PDF or CDF. An example is the paper [SW81].

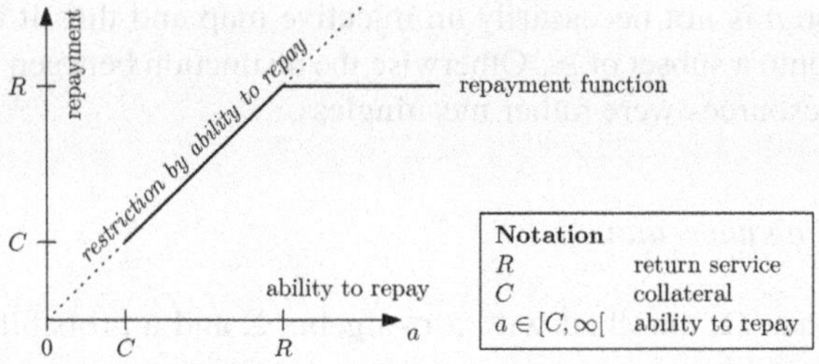

3.9 Collateral and the ability to repay
To illustrate that collateral is the lower bound of the ability to repay it is helpful to draw the repayment function in a cartesian plane with the vertical axis intersecting at zero. Compare figure 3.5. A repayment function specifies repayment for all *possible* situations (p. 35), hence the amount of collateral coincides with the lower bound of the repayment function's domain.

What about modelling collateral? Assets pledged as collateral are part of the resources at repayment time and thus a feature of the ability to repay. In the context of investment credits we have to distinguish *inside collateral* and *outside collateral*. Inside collateral refers to assets used in the investment project whilst outside collateral are assets not used in the project (see [Coc00, p. 192] or [CK85, p. 84]). If project success is modelled stochastically then inside collateral – as a part of this random variable – is naturally also modelled stochastically. In contrast, the value of assets used as (outside) collateral is most often modelled as a constant. However, there are some authors that consider *stochastic* (outside) collateral [Nii11, Bar76, JN02, FR15]. In this case the ability to repay has a positive lower bound (the value of the outside collateral) and this lower bound is not fixed but modelled stochastically.

Q 3.3 The value of collateral

Many things can serve as collateral, e.g. real estate like land, equipment like machinery, the inventory of a company or any other property or capital asset. Lending theory abstracts from these concrete assets used as collateral and concentrates on its value. What sounds trivial has some subtleties in store.

- Value to whom? The value of some item contemplated as collateral may be different for the debtor and the creditor. The difference may arise from subjective opinions and beliefs, e.g. a house may be more valuable to its long-time occupants than to a commercial bank. It can also arise from asymmetric information concerning the asset provided as collateral (see [CK85, p. 85]). Finally, costs of collateral, e.g. for marketing and selling the asset, may drive a gap between the lender's and borrower's valuation of the asset (see [Coc00, p.[192]).
- Value at what time? The value of collateral can be stochastic (see e.g. [Nii09]). That means that the ex ante expected value of the asset and the ex post realized value can differ.[a]

Thus, when working with collateral it is important to see the point! If collateral is used for screening, its expected value to the borrower is relevant. If it is used to limit the maximal loss of the lender, its expected value to the lender is relevant.

a 'Ex ante' vs. 'ex post' here refer to the realization of the random variable that models the value of collateral – not to the ability to repay.

3.3.4 Special case: Joint liability

General construction

How do we adapt the general construction of the ability to repay for the case of joint liability? First, we need a multivariate random variable (a random vector)

$$a = (a_1, \ldots, a_n)\colon (\Omega, \Sigma, P) \to (\mathbb{R}^n, \mathcal{B}(\mathbb{R}^n))$$

with n being the number of borrowers.[31] The component $a_i\colon \Omega \to \mathbb{R}$ represents borrower i's available resources for every possible state of the world in Ω. The pullback of P along $a = (a_1, \ldots, a_n)$ this time yields a multivariate probability distribution A, the joint distribution of the $a_1, \ldots, a_n$. The joint distribution of real-valued random variables can be specified by a joint PDF[32] or a joint CDF.

31 Side note: $\mathcal{B}(\mathbb{R}^n) = \bigotimes_{i=1}^n \mathcal{B}(\mathbb{R})$ the product-σ-algebra, as $\mathbb{R}$ is separable.
32 If existent.

In the literature

Which specifications of the ability to repay do we find in the joint liability literature? Concerning marginal distributions the situation seems reversed compared to the individual liability case above. In papers about joint liability the authors usually work with discrete distributions and probability mass functions rather than PDFs. Most of the time a simple Bernoulli-Distribution is assumed (e.g. in [Arm99, Gha00, BO10, KL12, RS13] and many more). Two rare exceptions are [BC95] with a continuous CDF to model the independent marginal distributions and [All16] who uses a general joint distribution.[33]

Concerning dependence structures – apart from very few publications like [Laf03, AT07, KL12, All16] – independence of the borrowers' ability to repay is the predominant assumption. The underlying reason for these overly simple probability distributions may be technical: The success on several independent Bernoulli experiments follows a Binomial distribution and is particularly easy to calculate.

Excursus: The issues of overly simple dependence structures

The remainder of this section broaches the issue of the overly simple dependence structures used in joint liability lending models in the literature. Is stochastic independence a realistic assumption? Unfortunately, it is not.

In reality stochastic dependencies naturally arise

- between similar investment projects, e.g. two dressmakers that are hit by a break-down of demand simultaneously,
- due to regional proximity, e.g. two farmers growing different crops that are both hit by a regional drought,
- because of common triggers, e.g. high energy prices that reduce the dressmaker's and the farmer's profit.

Thus the assumption of stochastic independence between the borrowers' ability to repay can lead to a massive underestimation of joint risks. Especially in the context of microfinance group lending, where it is not uncommon that all borrowers of a group live in the same village, independent project returns are an inadequate assumption due to regional proximity.

33 Note, however, that both papers then combine their continuous probability distributions with *discrete* repayment functions instead of applying *continuous* ones. See figure 3.8 on p. 39 for an example of a continuous repayment function with (partial) joint liability.

The focus on stochastic independence (and sometimes pairwise linear correlation) as dependence structure in lending models is therefore problematic! The problem is the applicability of these models: The majority of joint liability models is only valid for a very particular – and moreover very unrealistic – special case! Readers should not expect the results gained under independence to extend to realistic cases with dependence. In fact one can already observe how results become invalid and statements are no longer true for other dependence structures. An example: positive assortative matching may break down under correlation (see [KL12, p. 447]). Actually, a major task of future research will be to recalculate all results about joint liability lending that were obtained under the assumption of stochastic independence of the borrowers' ability to repay, to check their validity under realistic circumstances.

But be aware, simply using linear correlation as in [KL12] does not help us out here. If all joint liability models were recalculated with (linear pairwise) correlation, we would still neglect a wealth of other dependence structures. The reason: tail dependences! Particularly relevant for joint liability is the tail of a distribution, i.e. the likelihood that both borrowers end up with a low ability to repay simultaneously. There are dependence structures with high correlation but independent tails where, as McNeil et al. explain, "extreme events appear to occur independently in each margin" [MFE05, Example 5.32, p. 211]. Conversely, zero correlation does not imply independence and there are copulas with correlation 0 that exhibit tail dependence (see [MFE05, Table 5.1., p. 212]). Linear correlation is therefore an inappropriate tool (outside the class of elliptical distributions).[34]

Mathematics provides a concept to catch these tail dependencies: Copulas. If we want to analyse just joint liability, we can take a general multivariate probability distribution without bothering about dependence structures. The result will hold in general. If we want to compare joint liability to individual liability, however, a breakdown of the joint probability distribution into its margins and the dependence structure is inevitable, as for individual liability we need the individual margins without the dependence structure.

The mathematical foundation for this separation of marginal distributions and dependence structure is provided by Sklar's theorem. It states that each joint probability distribution can be composed from the marginal distributions plus a copula. If we

34 For a detailed discussion of linear correlation, its shortcomings and pitfalls, see [MFE05, Subsection 5.2.1, p. 201-206], or the article [EMS02].

denote the joint CDF of the probability measure A by F_A, and its marginals by F_{A_i} $(i = 1, \dots, n)$, then Sklar's theorem says that

$$F_A(a_1, \dots, a_n) = C_A(F_{A_1}(a_1), \dots, F_{A_n}(a_n))$$

for a copula C_A. (See [Nel06, Theorem 2.3.3] or [MFE05, Theorem 5.3].) These copulas are the appropriate mathematical tool to model reasonable dependence structures. But so far, lending theory has mainly focused on the independence copula $\Pi(u,v) = u \cdot v$ and ignored most of the rest.

3.4 Purpose enforcement

3.4.1 Two types of ex ante moral hazard

The ability to repay (p. 40) plays such an important role for the creditor (e.g. for determining the risk premium, see p. 34 f.) that he would like to fix this probability distribution right at contracting time. Nevertheless, the debtor is the one who has control and can influence his cash flows. This provides the breeding ground for ex ante moral hazard.

Definition (Ex ante moral hazard). *Ex ante moral hazard* refers to all actions of a borrower during the uncertainty phase that intentionally worsen his ability to repay, to the disadvantage of the lender, compared to contracting time.

Notice that the word 'worsen' in the definition above triggers a comparison, and to understand it we need two bits of information. Firstly, we need to know what is considered 'worse', i.e. the definition of the order. The definition above specifies 'worse' by the amendment 'to the disadvantage of the lender'.[35] Secondly, we need to know which two things we compare, i.e. the elements of the order that we pick for comparison. The definition above also specifies these by adding 'compared to contracting time'.

Now, should we expect ex ante moral hazard to be an issue or are the incentives of borrower and lender in line with each other, anyway? Unfortunately, they are not. Figure 3.10 shows the repayment function (compare p. 36), which is what the bank receives in different states of nature, and it depicts the outcome of the borrower, which is simply his resources a minus repayment. If his ability to repay is more than what he owes the creditor, the debtor can keep the rest for himself. At first glance it

35 What is detrimental to the lender is intentionally left open here. It depends on the evaluation criteria of the lender, for example his risk attitude. For details on evaluation, see section 3.6.

seems as if the borrower should strive for high realisations of a, which would be appreciated by the lender, too. But that is only true in the interval $[R-C,\infty[$. For realizations left of $R-C$ the borrower is indifferent. This flat part is the source of two types of ex ante moral hazard.

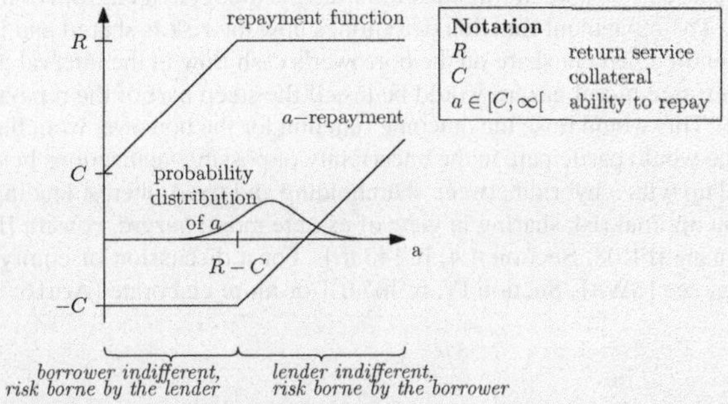

3.10 Fundamentally conflicting interests
The figure illustrates the fundamental conflict of interest between creditor and debtor. The primary vertical axis indicates the the value of flows whilst the secondary axis (not in the chart) indicates a probability.

Firstly, should it loom that the realization will most likely end up below $R-C$, then the borrower may give up all remaining efforts that could increase his ability to repay.[36] This will benefit his utility as he saves costly effort, but it will be hurtful to the bank. We have fundamentally conflicting interests and will most likely observe the ex ante moral hazard *low effort*.

And secondly, imagine the possibility of choosing another probability distribution for a with increased upside chances but at the costs of also increased downside risks. If the left tail of the distribution extends into the interval $[0,R-C[$, only the creditor is hurt by these realizations. Hence, both parties interests differ, the borrower profits of increased chances and the lender is harmed by increased downside risks. We have fundamentally conflicting interests again and will probably observe the ex ante moral *hazard diversion of funds*.

Altogether, we can keep hold of the fact that – concerning borrowers' incentives – the tail of the distribution extending into $]0,R-C]$ is problematic. The borrower has some influence on the probability distribution of a and he may either worsen the

36 This requires a model with gradually unfolding information. Gradually unfolding information can be formalised by structuring the sample space with a sequence of ever refined information partitions, see [MQ02, p. 215].

distribution in the course of the uncertainty phase (low effort) or choose a different distribution right away (diversion of funds).[37] Ex ante enforcement tries to prevent manipulations of a that are derogatory to the lender.

Q 3.4 Ex ante moral hazard – a risk sharing problem

The ultimate cause of ex ante moral hazard, in fact, is the repayment function of a standard debt contract. The repayment function determines how the risk is shared and it is the lender that holds a hundred percent share on the borrower's cash flow in the interval $[0, R-C]$. The incentive for ex ante moral hazard would be less if the steep part of the repayment function had slope < 1. This would raise the outcome function for the borrower from flat to a positive slope so that he would participate in the uncertainty of possible realizations below $R-C$, too. We would end up with a hybrid between shareholding and fixed-interest lending. For general information on optimal risk sharing in view of ex ante moral hazard, consult [Hol79, Inn90] or for a summary [FR08, Section 4.4, p. 143 ff.]. For a discussion of equity financing vs. debt financing, see [SW81, Section IV, p. 407 ff.] or more elaborate [Arn16, Section VII.6, p. 241 ff.].)

3.4.2 Monitoring games in the literature

Definition (Purpose enforcement). *Purpose enforcement*, comprises all means that help to prevent ex ante moral hazard (p. 46).

The borrower shall use the credit as stipulated to keep the quality of his ability to repay at its level at contracting time. As always with enforcement problems, the task consists of two separate challenges: The detection of moral hazard (⊠ of ④ on p. 16), and the punishment of the same (⚡ of ④ on p. 16).

My suggestion is to use the term 'monitoring' to refer to the informational aspects of purpose-enforc ement (see section 3.1.1). In the literature, however, we find the following definitions of monitoring:

- "Monitoring in this context means the application by the lender of some external control mechanism that makes the diversion of project resources towards private uses less profitable." [RS00, p. 1937]
- "monitoring enables the bank to force the entrepreneur to undertake the agreed-upon effort level" [BK93, p. 218]

Obviously, these definitions refer to different types of moral hazard, diversion of funds and low effort, respectively. Unfortunately, both definitions do not distinguish between the informational aspect and the incentive aspect of enforcement.

37 For a formal analysis of incentive effects, see [SW81].

In the literature authors use different approaches to incorporate monitoring into their models. [RS00] for example use a 'private benefit function' to represent a borrower's benefit as a function of his diligence. In their model, the case with monitoring is represented by a flattened private benefit function. [Con99] present a similar approach. A diligent borrower may divert a certain fraction of funds to gain some private benefits. His reward from diverting resources depends on a 'monitoring function' and is less for higher intensities of monitoring. In [HT97, p. 668], the borrower can choose between three different investments and "monitoring reduces the firm's opportunity cost of being diligent (by eliminating the high benefit B-project)" [HT97, p. 671]. The plain concept in all of these examples: Monitoring simply reduces the borrower's benefit from ex ante moral hazard.

A more elaborate possibility is to model the monitoring game explicitly. In [GG99, p. 205], a borrower privately decides upon his action (potentially a moral hazard). Then firstly, a monitoring intensity $m \in [0,1]$ is chosen. Secondly, the true action of the borrower can be observed with probability m and remains in the dark with probability $1-m$. And thirdly, in case moral hazard is uncovered, social sanctions are imposed. In such explicit monitoring games the two aspects of enforcement are modelled separately.

Q 3.5 Delegated monitoring

Since it takes up some space in the literature about monitoring, *delegated monitoring* should be briefly mentioned here. If several lenders jointly finance the same borrower it is in principle sufficient if one of the lender engages in monitoring and creates a sufficiently large incentive effect against ex ante moral hazard. Most frequently this role is taken by financial intermediaries which provide monitoring services. The other lenders simply benefit from the monitoring work already being done.

Lots of literature about monitoring appears in conjunction with the question why intermediaries and markets coexists (see for example [Dia84, BK93, RS00, Par00]). One explanation: Monitoring can be achieved more effectively by banks than markets! A list of reasons why that is the case can be found in [RS00, p. 1937, 1938]. Thus one reason for the existence of banks may be: It's them performing all the monitoring work.

3.4.3 General structure of monitoring games

When modelling monitoring games, some questions arise. One issue is, what happens first, the private action or the choice of monitoring intensity? If monitoring intensity is fixed first and observed by the borrower, this may influence his private decision. Another point is related to the costs of monitoring. With costless monitoring the highest level of monitoring is attractive. With rising costs monitoring becomes charmless. Thus the equilibrium is highly dependent on how the costs of monitoring are modelled.

In principle, however, the idea is always the same: Within the monitoring game, the borrower must balance out the benefit of moral hazard against the risk of being uncovered and sanctioned, whereas the lender must trade off the disciplining effect of monitoring against the costs of monitoring. We get an equilibrium by searching for each others' best response. The equilibrium depends on the calibration of the strength of each effect.

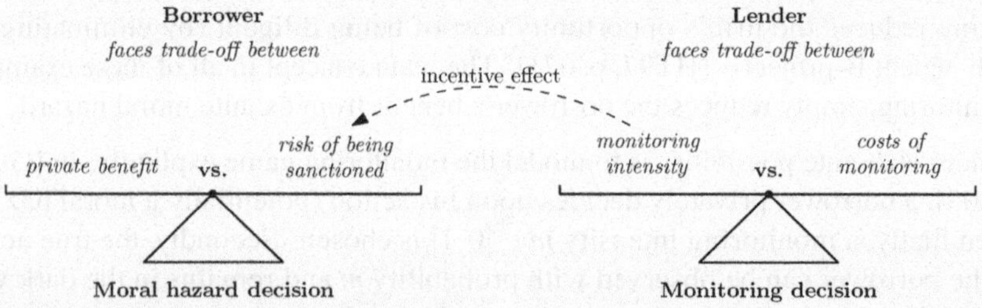

3.11 Monitoring game structure
This figure illustrates the common structure of monitoring games. Players are the borrower and the lender. Their action sets formalize the moral hazard decision and the monitoring decision, respectively. The interdependencies between the players actions and their payoffs are characterised by, firstly, two trade-offs, and secondly, the influence of the monitoring intensity on the risk of being uncovered. Concrete repayment models vary concerning the timing of the model, discrete vs. continuous action spaces and the concrete specification of the payoff functions.

3.4.4 Special case: Group lending contracts

There exist several publications on ex ante moral hazard in conjunction with group lending e.g. [Sti90], [GG99, Section 2.2], [Con99, Con05, Zha08]. For a well-written literature overview, see [Zha08].

Peer monitoring – when members of a group credit monitor each other – is actually quite similar to the idea of delegated monitoring. It is sufficient if only *one* stake-holder pays attention to ex ante moral hazard. Why not let the borrowers monitor one another and save the bank the monitoring effort?

The idea suggest itself to ask for the enforcement capabilities of borrowers. Again, we have to consider the two components of enforcement: observation and punishment (p. 24). Many models in the microfinance literature assume that group members in a group credit have an informational advantage compared to the bank. Maybe they live close to each other or they know each other for a long time already, such that monitoring is less costly for them than for the bank. But beyond better information, it may also be the ability to set strong incentives that constitutes the borrowers advantages. Social sanctions may be stronger than any threat uttered by the bank. Summing up, there are reasons to assume that borrowers have effective and efficient peer-monitoring and peer-sanctioning qualities, and hence enforcement capabilities.

The next question associated with group monitoring is whether the borrowers have an incentive to monitor. Or do we end up with another instance of moral hazard, where borrowers go easy on their peers instead of monitoring them as stipulated? Let's go through the three forms of group lending presented on p. 31. Under joint liability we can expect that each group member is interested in the other group members being able to repay their loan, as their own outcome is directly linked to their peers repayment. Thus, they have an incentive to monitor each other. Likewise with sequential group lending. Under pure cross-reporting without joint liability, however, the question does not have an obvious answer, collusion may be a problem.

Nevertheless, in the cases of joint liability and sequential lending, the contract inherent incentive to monitor your peers can be seen as a potential third advantage of peer monitoring (in addition to the potentially superior monitoring and sanctioning efficiency between peers). [Con05] compares delegated monitoring and peer monitoring, but despite other monitoring models in his paper peers are not equipped with a superior monitoring or sanctioning technology. In his analysis the advantage of joint liability rests on an "incentive diversification effect that cannot be replicated by outside delegates or intermediaries." [Con05, p. 1] In a nutshell, joint liability reduces the costs of setting incentives. He explicates: "Where the total cost of lending under a delegated arrangement would be the sum of limited liability and delegation rents, a single rent suffices to provide incentives in the joint-liability multi-task setting." [Con05, p. 5]

In addition to the subtleties of bank monitoring like timing and the calibration of costs and effects (see p. 49), the modelling of peer monitoring has even more intricacies in store. Obiously, one issue concerns the degree of cooperation amongst group members. Do they make their effort decision cooperatively or non-cooperatively? The difference matters, see [Zha08]. In addition, an even more basic aspect flashes up: With more than two borrowers the question 'Who monitors whom' does not have a unique answer anymore. Different monitoring structures have been analysed by [Arm99]. Figure 3.12 shows three examples. Furthermore, Armendáriz mentiones that – depending on the monitoring structure – pair-wise collusion may become a problem [Arm99, p. 99]. Altogether, the formalisation of a peer monitoring setting is decidedly more complex than modelling monitoring by the lender.

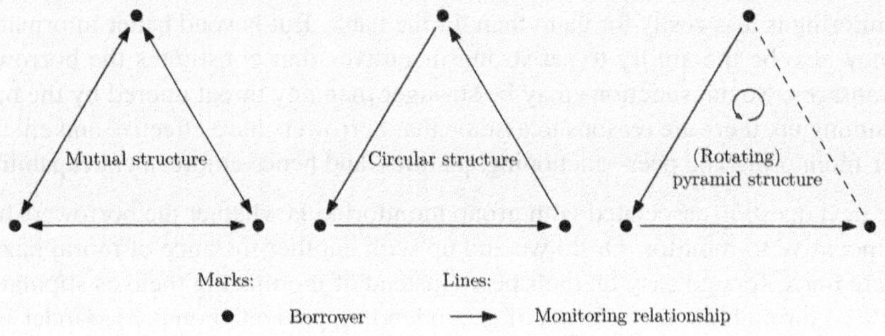

3.12 Peer monitoring structures
The chart illustrates the three different monitoring structures mentioned in [Arm99, Section 5]. The rotating pyramid structure requires several periods where borrowers take turns in monitoring their peers.

3.6 Quantity of monitoring structures

Of course there are much more possible configurations than the ones presented by [Arm99] and depicted in 3.12 – especially when a larger number of borrowers is involved. A rigorous analysis of peer-monitoring structures starts with an examination of the set of possible configurations. Borrowing terminology from graph theory, a graph representing a monitoring structure should satisfy the following properties:

Labelled The vertices must be labelled, otherwise the rotating pyramid cannot rotate.

Directed Edges must be directed to specify who monitors whom.

Loopless The graph must be loopless as there is no point in 'self-monitoring'.

Let's use combinatorics to get an impression of how large the set of possible monitoring structures is. Given n borrowers, the number of labelled, directed, loopless graphs with no repeated edges is $2^{n(n-1)}$, which is 1.048.576 for $n = 5$.[a]
If we additionally require the indegree of each vertice to be exactly 1 – the most efficient monitoring where each borrower is surveilled by exactly one peer – the number of configurations is still $(n-1)^n$, giving 1.024 for a group of 5.[b] In addition the edges may need to be weighted to account for the fact that different borrowers monitor variably well.
Of course, with labelled edges, the circular structure comes in $\frac{n!}{n} = (n-1)!$ variants, which might be considered conceptually the same.[c] So we need to define an equivalence relation on the set of monitoring structures.
Altogether, the comprehensive analysis of monitoring structures poses a challenging task for future research.

---

a Each potential edge can be turned 'on' or 'off'.
b A monitoring peer must be chosen for each of the n borrowers.
c Determining the path is equivalent to drawing n without returning, the denominator accounts for permutations.

3.5 Payback enforcement

In this section we tackle the most obvious problem of deferred reciprocity agreements: How to ensure that the debtor delivers the promised return service? First of all we ought to precisely define the problem.

3.5.1 Definition of strategic default

Definition (Strategic default). *Strategic default* refers to an action of a borrower where he repays less than the minimum of the realisation of his ability to repay and the stipulated return service.

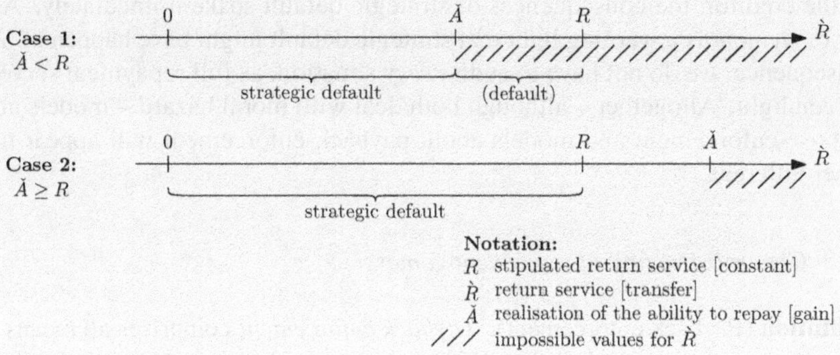

3.13 Actions labelled as strategic default
According to the definition, which values of $\grave{R}$ get labeled as strategic default?

The definition above is a precise but rather formal way to say what strategic default is. Equivalently, we could define: Strategic default refers to an action where the borrower intentionally reduces the value of the return service compared to the stipulated amount.[38] In this alternative formulation the definition resembles that of ex ante moral hazard on p. 46 – which is no surprise as both are instances of a moral hazard (see p. 149).

38 To recognise the equivalence, see figure 3.13. A formal proof would have to distinguish the two cases depicted there.

So can we treat strategic default the way we handled ex ante moral hazard? Unfortunately not. There are distinctive differences between ex ante moral hazard and strategic default. Firstly, concerning the thing controlled: For ex ante moral hazard the borrower has control over the form of a probability distribution (by manipulating the random variable ability to repay, p. 40), whilst for strategic default the borrower has control over concrete resources and the apportionment of the same between lender and borrower. Secondly, which actions of the debtor are considered to be a moral hazard is not constant in models about strategic default. Instead it changes with the realisation of the ability to repay (see figure 3.13). Thirdly, looking at the information asymmetry involved: Ex ante moral hazard is often based on hidden action (the manipulation itself remains unobserved by the lender), whereas strategic default is mostly based on hidden states (the repayment decision itself is not hidden at all, instead the borrowers repayment capability is unknown).[39] And lastly, whilst ex ante moral hazard can remain undiscovered without instantaneous consequences for the creditor, the consequences of strategic default strike immediately. A too low repayment is a warning light that strategic default might have happened. As a consequence, we do not have to audit *every* situation, as full repayment serves as a green light. Altogether – although both deal with moral hazard – models about purpose-enforcement and models about payback-enforcement will appear to be rather different.

3.5.2 General structure of repayment games

Definition (Payback enforcement). *Payback enforcement* comprises all means that help to prevent strategic default (p. 53).

Again, the enforcement problem splits into two challenges, namely the detection of strategic default (⌧ of ③ on p. 16) and the sanctioning of the same (↯ of ③ on p. 16). The term *auditing* refers to the informational aspect of payback enforcement.

The tool to model and analyse the repayment stage of a credit contract are *repayment games*. They answer the fundamental question: How will the borrower behave at repayment time? We will follow the PAPI-structure (players, actions, payoffs, information; see [Ras07]) to describe the basic structure of such games.

39 Usually it is that way. However, we can imagine a borrower that observably exerts low effort but it remains unclear whether he simply does not or cannot do more; then ex ante moral hazard would covered by hidden states (about the borrower's capabilities). Conversely, there are instances of strategic default that are based on hidden action, e.g. if a borrower intentionally but unobservably worsens the value of his collateral, knowing that he will have to transfer it to the lender soon.

Players: The single player of a repayment game is the borrower.

Actions: His action set consists of the different repayments he can make. It is restricted by the realisation of his ability to repay. Sometimes the action space is enriched by additional features, e.g. messages the borrower can send.[40]

Payoffs: The borrower's payoff is composed of three components.

- Resources: the amount of resources he calls his own at repayment
- Repayment: the amount of resources he transfers to the lender
- Penalties: the punishment imposed by the lender

Information: Possibly hidden information about the borrower's available resources (hidden states, see p. 26) restrict the accuracy with which the lender can target his sanctions.

The following explanations elaborate on some details of this structure.

Why is the lender not part of the repayment game? It is tempting to think of the penalties as an action of the lender. But, the actual decision which sanctions are imposed in which situation is made long before repayment time. The sanctions are typically fixed in the credit contract. So although penalties come into effect not until the repayment game, the decision is made at contracting time and thus belongs to the contracting game.[41]

How are different realisations for the ability to repay considered in the repayment game? Indeed, repayment games come in *bundles*, that is, we have an individual repayment game for every realisation of the ability to repay. The individual members of a bundle differ in the configuration of the borrower's action set – with more resources he has a wider range of choices. Since the borrower's resources 'shape' his action set, I refer to the individual members of a bundle as different *shapes*.[42] To analyse the overall performance of a repayment game we have to take into account how it performs in every shape and weight the individual equilibrium outcomes with the appropriate probabilities.

What do the payoff components depend on? The *resources-component* depends on the shape of the repayment game but its magnitude is invariant within a shape. Therefore within a shape this component, e.g. the yields of an investment project, are irrelevant for the calculation of the equilibrium. The *repayment-component* of the payoff obviously depends on the borrowers repayment, i.e. the action he chooses.

40 Notice that the amount repaid can itself be seen as a message.

41 In this monograph we focus on contracts that are non re-negotiable. Also see p. 59, box 3.8.

42 Imagine the whole lending game as a big extensive-form game tree. Then the shapes are the subgames that hang on the leaves of nature's ability-to-repay-decision-point.

The more he repays the less his payoff will be. Finally, the *sanctions-component* also depends on the borrower's repayment but, if the lender can get hold of the information on the realisation of the borrower's ability to repay, it can additionally be made dependent on the borrower's resources at repayment time. In such case, the punishment function varies from shape to shape. Notice that the punishment rule of a repayment game only specifies the conditions and intensities of sanctions. The means by which the penalties are implemented are treated as a blackbox here (see sections 2.4.2 and 3.1.2).

How does the information distribution affect the repayment game? It determines the structure of the domain of the punishment rule (also called penalty function). Given hidden states, the domain of the punishment rule can only be the action set of the borrower. With full information, domain may be the Cartesian product of the borrower's action set and the image of the ability to repay.

How is the penalty function connected to the repayment function? The repayment function specifies the stipulated return service in various situations (p. 35). If the actual repayment deviates from the amount specified, sanctions are imposed, i.e. the penalty function is based on the contract's repayment function. Of course, the purpose of these sanctions is to prevent deviations from the repayment function. Thus we can view the whole repayment game as the machinery that implements the repayment function of the credit contract. If the repayment function is enforced by the repayment game we call the contract *incentive-compatible*. In other words: A contract is incentive-compatible if repaying as stipulated is a dominant strategy.[43]

3.5.3 Repayment games in the literature

In the literature incentive-compatible contracts have been studied under various assumptions concerning information asymmetries (⊠) and punishment capabilities (↯).

⊠: *hidden states* | ↯: *unlimited* In [Dia84, p. 396] the realisation of the debtor's ability to repay is unknown to the lender. Concerning punishment capabilities the creditor is equipped with the possibility to exert arbitrarily large penalties. The penalty function is designed to exactly offset the difference between the actual repayment and the stipulated return service. In his setting the optimal repayment function, i.e. the one that can be enforced with minimal expected non-pecuniary costs, is that of a standard debt contract (see [FR08, p. 135]).

43 Other authors define incentive compatible contracts the same way, e.g. [Tow79, p. 266], [FR08, p. 130] or very formally [GH85, p. 653].

⊠: *hidden states* | ↯: *limited* Some papers examine the denial of future loans as a punishment, e.g. [BS90, Gro94] or for sovereign debtors [FR08, Section 4.3.4]. This is, of course, an implementation topic and actually belongs to the incentive tools sections (p. 27). Nevertheless, these models are also interesting from a repayment game perspective since this assumption implicitly says that the intensity of the penalty is restricted to the value of future loans. So compared to [Dia84], these are models with hidden states and limited punishment capabilities. Notice that in repayment games with hidden states the penalty function cannot distinguish between accidental default and strategic default. It can only be contingent upon the observable repayment decision of the borrower. The challenge in such situations often is to set the intensity of punishment to a level that trades off its positive incentive effects (preventing strategic default) with its negative welfare effects (hitting unsuccessful borrowers).

⊠: *full information* | ↯: *unlimited* Given full information and unrestricted punishment capabilities enforcement is trivial.

⊠: *full information* | ↯: *limited* In [JPB05] "Banks can observe whether projects succeed or fail, so that there is no asymmetric information" [JPB05, p. 225]. The repayment function requires the borrower to repay the agreed payment $1+r$ in case of success and $\min\{(1+r),\phi_c c\}$ in case of failure ($\phi_c c$ is the amount of retrievable outside collateral, ϕ_c the recovery rate). In case of strategic default the penalty function deducts $\phi_p(1+\pi)+\phi_c c$ from the borrower's payoff (ϕ_p being the recovery rate of the project yield $(1+\pi)$). Obviously, the contract is incentive compatible if and only if $1+r \leq \phi_p(1+\pi)+\phi_c c$. The paper is an illustrating example how it is sometimes difficult to see behind all the notation and interpretation in order to extract the essential structure of the repayment game. The punishment implementation (see p. 21) used in [JPB05] is judicial enforcement. However, the essence behind all the story is a repayment game with full information and a punishment capability that is limited by $\phi_p(1+\pi)+\phi_c c$.

Finally, there are models that lie between full information and hidden states. If states are in principle hidden but the lender has access to a costly auditing technique to find out the true state of the borrower's ability to repay, we speak of *costly state verification*. The auditing must be expensive, otherwise it would be just the same as full information. To save auditing costs an *auditing rule* is introduced, specifying for which actions of the borrower an audit is executed. For example, if the borrower fully repays, auditing is needless. Notice that auditing neither changes the basic structure of the repayment game nor does it add another decision point. The auditing rule (like the penalty function, see p. 55) is defined at the contracting stage. Its influence on the repayment games comes by determining the information distribution in certain situations. In turn, this determines what the penalty function can be contingent on. [GH85] present a model with costly state verification. Their approach is quite general. They take an arbitrary repayment function $C_1(s)$ that depends on the state s of the ability to repay, an auditing rule $B(.)$, costs of auditing $c_1(.)$ and a discrete penalty function with values c_0 or zero, respectively (see [GH85, p. 651, 652]). After characterising all incentive compatible contracts (proposition 1), they chase the contract with minimal auditing costs and penalties and arrive at a standard debt contract (proposition 3).[44]

As with monitoring games (see section 3.4.3) there are some subtleties in repayment games, too. Firstly, the outcome is dependent on how the costs are modelled. In the repayment game just sketched, the costs of auditing $c_1(s,l)$ depend on the state s and the size of the investment l (see [GH85, p. 651]). But as Townsend points out, these can be modelled in various different ways: "One natural specification is that the cost of verifying y_2 [the realization of the random variable] is some constant ... independent of the actual realization; ... The cost also may be supposed to depend on y_2, either directly or, alternatively, through the agreed-upon transfer. ... One may argue, for example, that the cost of auditing a firm in bankruptcy proceedings depends on outstanding claims." [Tow79, p. 269]

Secondly, the ex ante defined auditing rule must be credible. The lender may threaten the borrower with an auditing rule to create an incentive to repay, but eventually – since auditing is costly – forgo the actual audit. Concerning the credibility of auditing rules, [KP98] point at loan size, which may play a crucial role in convincing the debtor that the creditor has a high stake in an audit.

Thirdly, a topic that is discussed in the literature is stochastic auditing rules. There, the audit is executed with a certain probability only. Such contracts can be imagined with a built in forgiveness; some defaulting borrowers get away without being audited. The advantage? By auditing the borrowers' selectively the lender can save audit costs. However, if the probability of an audit is too low, the incentive effect is gone. [MP89], for example, extend the costly state verification model with deterministic auditing of [Tow79] to stochastic auditing and search for the optimal stochastic audit strategy.[45] [BS94, p. 540, 541] provide a detailed list of important literature about costly state verification with deterministic and stochastic auditing.

44 Actually, their argumentation takes a little detour. First, they jump into the evaluation module of lending theory (see section 3.6) and define what they consider as optimal, namely contracts that "maximise the entrepreneur's expected utility subject to the investor's zero-profit condition." [GH85, p. 653] Second, they explicate that the pecuniary costs of auditing $c(.)$ are effectively borne by the lender like the non-pecuniary costs c_0 are (see [GH85, p. 651 f.]). Thus they are simply searching for the contract that minimises expected costs of auditing and the penalties.

45 Both papers argue in a general principal-agent framework, but their arguments are applicable to the lender-borrower relationship, too.

Q 3.7 The popularity of standard debt contracts

Why are standard debt contracts so common if stochastic auditing provides a more efficient alternative? An interesting paper concerning this question is penned by Boyd and Smith [BS94]. They compare the cost efficient stochastic auditing to optimal deterministic auditing (a standard debt contract). Using a calibrated model they arrive at the conclusion that the welfare loss from staying with less efficient standard debt contracts is "extremely small" and that "standard debt contracts will be (very close to) optimal." [BS94, p. 539].

Q 3.8 Renegotiating the contract after default

Since it takes some space in the literature, the incomplete contract approach should be mentioned here. As Freixas and Rochet explain, it might be to difficult to ex ante describe all the events that can occur and write an extensive contract that explicitely deals with every single one [FR08, p. 146]. Incomplete contract theory addresses the problem. For an illustrating example see [FR08, Section 4.5], and for a synthesis of the incomplete contracts literature see [Jea03] or the book [BD05, Chapter 12].
Other authors also leave some slack in the credit contract by making the punishment decision a decision node in the timeline instead of hard-coding the penalty function into the contract at contracting time (see for example [KV00, Sha03]).

3.5.4 Special case: Group lending

When it comes to group lending things change a little bit. Let's go through the differences. Firstly, concerning players and actions, we have more than one borrower and accordingly more than one action set. Secondly, since each borrower has its own ability to repay the number of shapes of the repayment game increases. Thirdly, with respect to payoffs, the borrowers can impose sanctions on each other – so-called social sanctions – which adds an additional component to the borrowers' payoffs. Moreover, the penalty function gets more complex as it must specify the penalties for different borrowers in each situation. Finally, looking at the information distribution within the game, the question arises if borrowers make their decisions sequentially or simultaneously (i.e. with or without knowing the other borrower's repayment decision). Furthermore, we have to specify how informed they are about each other's realisation of the ability to repay.[46]

The following series of figures illustrates the basic structure of a group lending repayment game. The first table 3.14 is an example for a *penalty function*, also called *punishment strategy*. Depending on the borrowers' actions, the lender punishes one or both of them. In the example the punishment intensity is $-M$.

46 In the group lending literature the creditor is usually assumed to suffer from hidden states concerning the borrowers' ability to repay, whilst the borrowers themselves know each other's realisations.

	P2:0R	P2:1R	P2:2R
P1:0R	-M | -M	-M | 0	0 | 0
P1:1R	0 | -M	0 | 0	0 | 0
P1:2R	0 | 0	0 | 0	0 | 0

3.14 Punishment strategy (example)
Example of a punishment strategy of a lender. The punishment here only depends on the borrowers' repayments. (P1: action of player 1, P2: action of player 2. 0R,1R,2R = repayment decision. M = punishment intensity.)

The next table 3.15 shows the unrestricted shape of a group lending repayment game. The payoff in this example has two components, the amount repaid and the penalty. Recall that the realisation of the ability to repay – actually an additional component of the borrowers' payoff – is negligible for the equilibrium of a shape, as it is constant within each shape. Notice that 0R,1R,2R represent the elements of the action set as well as the amount repaid, i.e. they serve as a label and as a number. For repayment games in combination with a message game the labels for the elements of an action set get more complex (e.g. see figure 12.4 on p. 185).

	P2:0R	P2:1R	P2:2R
P1:0R	-M | -M	-M | -R	0 | -2 R
P1:1R	-R | -M	-R | -R	-R | -2 R
P1:2R	-2 R | 0	-2 R | -R	-2 R | -2 R

3.15 Payoffs in an unrestricted shape of a repayment game (example)
Example for borrower payoffs for each repayment action combination in the unrestricted shape of a repayment game. (P1: action of player 1, P2: action of player 2. 0R,1R,2R = repayment decision and amount repaid. M = punishment intensity.)

Table 3.16 illustrates how the realisations of the ability to repay restrict the borrowers' action sets and form the shapes (p. 55) of the repayment game. Infeasible actions are marked with ×. To determine the performance of a repayment game in different situations, each shape must be solved separately. Therefore in each shape of the repayment game the best responses are underlined. There is a Nash equilibrium where two of them meet, i.e. the players play best responses simultaneously.

		A2:0			A2:-R			A2:-2 R		
		P2:0R	P2:1R	P2:2R	P2:0R	P2:1R	P2:2R	P2:0R	P2:1R	P2:2R
A1:0	P1:0R	[-M,-M]	×	×	[-M,-M]	[-M,-R]	×	[-M,-M]	[-M,-R]	[0,-2 R]
	P1:1R	×	×	×	×	×	×	×	×	×
	P1:2R	×	×	×	×	×	×	×	×	×
		P2:0R	P2:1R	P2:2R	P2:0R	P2:1R	P2:2R	P2:0R	P2:1R	P2:2R
A1:-R	P1:0R	[-M,-M]	×	×	[-M,-M]	[-M,-R]	×	[-M,-M]	[-M,-R]	[0,-2 R]
	P1:1R	[-R,-M]	×	×	[-R,-M]	[-R,-R]	×	[-R,-M]	[-R,-R]	[-R,-2 R]
	P1:2R	×	×	×	×	×	×	×	×	×
		P2:0R	P2:1R	P2:2R	P2:0R	P2:1R	P2:2R	P2:0R	P2:1R	P2:2R
A1:-2 R	P1:0R	[-M,-M]	×	×	[-M,-M]	[-M,-R]	×	[-M,-M]	[-M,-R]	[0,-2 R]
	P1:1R	[-R,-M]	×	×	[-R,-M]	[-R,-R]	×	[-R,-M]	[-R,-R]	[-R,-2 R]
	P1:2R	[-2 R,0]	×	×	[-2 R,0]	[-2 R,-R]	×	[-2 R,0]	[-2 R,-R]	[-2 R,-2 R]

3.16 Shapes of a repayment game (example)
Example of a repayment game with all shapes depicted. (A1: ability of player 1, A2: ability of player 2. P1: action of player 1, P2: action of player 2. 0R,1R,2R = repayment decision and amount repaid. M = punishment intensity. x = infeasible action. Underlined = best response.)

The details of the game can then be put to the background and the result can compactly be displayed by, what I call, a *map of equilibria* – or more precisely a *map of equilibrium actions*.

	A2:0	A2:-R	A2:-2 R
A1:0	0R | 0R	0R | 1R	0R | 1R
A1:-R	1R | 0R	1R | 1R	1R | 1R
A1:-2 R	1R | 0R	1R | 1R	1R | 1R

3.17 Map of equilibria (example)
This is an example that shows how a map of equilibria may look like. (A1: ability of player 1, A2: ability of player 2. 0R,1R,2R = equilibrium repayment decision.)

From the map of equilibria we can derive the map of equilibrium lender payoff, the map of equilibrium borrower payoff and the map of equilibrium dead-weight loss. And given the probability of each situation we can calculate the exact probability distributions for lender payoff, borrower payoff and dead-weight loss. And in turn, of course, also expected values of these distributions. It is not necessary to to keep track of all the details of the game. We will see in chapter 4 that the result of a repayment game is totally sufficient to proceed to further modules. The easy interface that connects a repayment game to the other modules of lending theory are the maps of equilibrium outcome for each player. The mechanics of a repayment game can remain in a black box.

3.9 Map of equilibrium repayment vs. repayment function
The map of equilibrium repayment as well as a repayment function, both map states to repayment. The essential difference: Behind a map of equilibrium repayment there is a whole enforcement game, whilst a repayment function is just a contractual request. Precisely stated the challenge of payback-enforcement is to construct a repayment game whose map of equilibrium repayment coincides with the repayment function. (See incentive-compatibility on p. 56.)

Of course, the game above is only one of many examples of repayment games. Concerning the borrowers' choices we can distinguish discrete action sets (as above) and continuous action sets – although in the microfinance literature discrete modelling predominates. The timing can of course be changed from simultaneous repayment decisions (as above) to sequential actions. And the punishment strategy can come in many different forms; for example it can be a 2-valued function (as above) that knows either punishment or no punishment, or n-valued with several shades of punishment intensity, or even continuous. Let's turn to the literature and have a look at some specialities of the group lending repayment games we find there.

Topic: Timing

Apparently, we can imagine simultaneous repayment and sequential repayment. In the literature simultaneous repayment prevails. But the literature also provides a third timing alternative where the repayment decision is separated from the bail-out decision. One example is the joint liability game of Besley and Coate (1995). The borrowers first of all decide simultaneously if they want to contribute their repayment share. But then, Besley and Coate add an additional stage to the game where – under certain circumstances – one borrower is given the chance to ramp repayment up to the full amount, see [BC95, Fig.1.]. Clearly, the bail-out decision happens in the second stage. Another example is the paper of Allen (2016), but with timing reversed. Here transfer payments are sent before the borrowers (simultaneously) decide whether to repay or not. So the decision whether to help a peer is made previous to the actual repayment decision, see [All16, Fig.1.].

Topic: Action sets

Under individual lending, whether the borrower simply repays or sends a message how much he can repay, is equivalent. The repayment can be interpreted as a message about the realisation of his ability to repay. Under joint liability, however, if the borrowers enjoy an informational lead and know each other's realisations, then the report of a borrower can contain valuable information. Rai and Sjöström present a model with an enriched action set: The borrowers repay and simultaneously release information about the resources of their peer. A verbal description of this cross-reporting game can be found in footnote 11 of [RS13, p. 259]. The advantage: Although the lender suffers from hidden states and cannot distinguish accidental default and strategic default, there is no punishment in the situation where both borrowers are unable to repay. And moreover Rai and Sjöström emphasise: "no agent is ever punished in equilibrium." [RS13, p. 259] Since punishment is a loss (see p. 103), this game holds tremendous welfare benefits. The drawback: The game is vulnerable to collusion. And if two borrowers can collude in this game, repayment never occurs. If they can collude only imperfectly, [RS04] show that a message game can improve efficiency.

Topic: Payoff components

Whilst Rai and Sjöström try to use the borrowers' information advantage to increase repayment rates, Besley and Coate exploit the borrowers' sanctioning capabilities. In the second half of their paper [BC95] they add another component to the borrowers' payoff functions, called social sanctions. With social sanctions free-riding equilibria disappear. With respect to repayment rates, there is one case in which the loan is repaid with social sanctions whilst it is not repaid without social sanctions. However, this superiority of repayment rates with social sanctions stands on shaky ground. They encounter multiple equilibria and deal with the ambiguity of equilibrium repayment by simply assuming that "repayment occurs in the ambiguous cases" [BC95, p. 12] (also see p. 65).

Topic: Repayment rates

Let's finally look at an interesting effect on repayment rates in group lending repayment games. Sometimes group lending repayment games exhibit shapes with less repayment than under individual lending. This is called the *perverse effect of joint liability* and origins from two reasons. Firstly, the realisation of the ability to repay might end up somewhere between $1R$ and $2R$, so given the peer suffers accidental default a complete bail-out is impossible. Since the two borrowers will default as a group anyway, they might decide to repay nothing at all even though there were some resources that could have been repaid. Secondly, if the maximum punishment intensity is too low, strategic default may simply be more attractive than a possible complete bail-out. Against that background the question arises of how to avoid this perverse effect of joint liability. The general direction seems to be making the joint liability part of the contract more flexible to avoid strategic default in the shapes that exhibit the perverse effect. [All16], for example, presents a contract with partial joint liability where optimal transfer payments are determined endogenously. The paper [DQFG16] suggests switching from explicit joint liability to implicit joint liability where a bail-out is only expected if resources suffice. In chapter 10 we will discuss another idea to make the perverse effect of joint liability vanish, namely making the punishment function more flexible.

Excursus: The treatment of multiple equilibria

A frequent problem in the group lending literature is the appearance of multiple equilibria. Two issues: Firstly, Basar and Olsder explain that "since the players make their decisions independently, it could so happen that the outcome of their joint (but non-cooperative) choices is not an equilibrium point at all" [BO99, p. 170], a so-called *coordination failure*. Secondly, even if the borrowers manage to coordinate themselves to play only Nash equilibria, it is unclear which equilibrium they play.

A possibility to at least ensure that an equilibrium strategy combination is played, is to assume that the players have access to some (unspecified) coordination mechanism. As long as all multiple equilibria yield the same overall repayment it is sometimes not necessary for the lender to know which equilibrium is played. But in *ambiguous cases*, where different equilibria exhibit different amounts of repayment, the situation is particularly problematic, especially with regard to calculating repayment rates.

How do authors handle the problem of ambiguous cases? It seems to be a widespread approach to simply add additional assumptions until a single equilibrium remains. A recent example is the paper [All16]. Allen writes: "Such an equilibrium is undesirable both theoretically ... and empirically ..., so I focus on the equilibrium where the loan is repaid if both group members are able to repay." ([All16, p. 203]) De Quidt et al. state that "although there may be many equilibria of the repayment game ... we focus on the welfare maximizing one." [DQFG16, p. 220] Another famous example is [BC95]. Confronted with two equilibria "mutual contribution" and "mutual non-contribution" they simply "assume that the first one of these is reached" [BC95, p. 7]. In the case with social sanctions an additional ambiguous case appears and since they assume repayment in ambiguous cases they arrive at the superiority of group to individual liability lending in terms of repayment rates (see [BC95, p. 12]). Imagine they had assumed default is played in ambiguous cases – their conclusion would be upside down. In the last line on page 7 the authors admit that "the fact that we are assuming mutual contribution for this case makes our calculation of the repayment rate only an upper bound on the performance of group lending" ([BC95, p. 7]), thus their analysis may overstate the performance of group lending and lead to inappropriate recommendations for practitioners.

I suggest that instead of simply neglecting undesired equilibria away, they must be taken into account! Therefore in the repayment games in the appendix all Nash equilibria are clearly presented. Due to the reasons explained above, multiple equilibria and ambiguous cases should be regarded as a defect in the design of a repayment game. Repayment games with multiple equilibria should be improved before applying them in reality.

3.6 Evaluation of credit contracts

In credit contract theory most papers consist of a comparison. What is the best repayment function under costly state verification (e.g. [GH85])? Is individual liability better than joint liability (e.g. [BC95])? Is explicit joint liability better than privately side-contracting borrowers (e.g. [RS13])? What is the best degree of joint liability (e.g. [All16])? Basically we always compare contracts. Therefore it is of vital importance to put the standards of comparison up for discussion. What do we consider as 'better'?

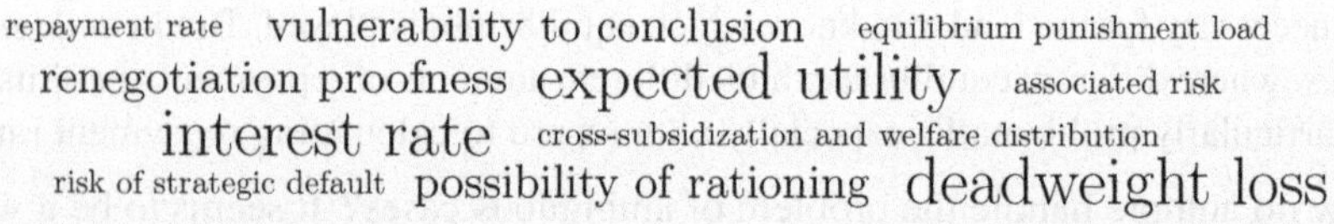

3.18 Word cloud of evaluation criteria
There is a large number of possible assessment criteria for credit contracts. Therefore a systematic approach to the evaluation of credit contracts becomes necessary.

3.6.1 A systematic approach to evaluation

On p. 18 we learned that we can abstract from the details of a credit contract and regard it as a vector of probability distributions of payoff. This vector has as entries one probability distribution of payoff for each borrower, one for the lender, and one for an artificial player that collects dead-weight losses.[47] *Evaluation of credit contracts* means judging this vector of probability distributions of payoff and enunciating preferences. Notice that the comparison of contracts 'lives' on the product layer of lending theory. We are interested in the outcome of a contract rather than the technical details of its implementation.

47 Dead-weight costs (e.g. cost of monitoring, dead-weight punishments) are borne by either the lender or the borrower, so they appear in the lender's and borrower's probability distribution of payoff, too. The collection of these costs in a separate distribution, without reference to the side on which they accrue, – although redundant – is useful when analysing the welfare impact of a credit contract.

When comparing credit contracts we can assume three different perspectives:[48]

The lender's perspective e.g. which contract yields maximal expected profit

The borrower's perspective e.g. which contract is the least expensive

The welfare perspective i.e. which contract is the most efficient

The first two perspectives are particularly important in the contracting phase. Depending on their evaluation of a particular contract, lenders and borrowers may accept or refuse it. The third perspective measures the performance of a contract. It is the perspective of the artificial player that collects losses (cost of collateral, cost for monitoring and auditing, dead-weight punishments for enforcement, deterioration of social relationships, etc.).

After choosing a perspective, we have to fix the criteria for comparison. Is the goal utility maximization or just satisficing[49]? Even if the goal is utility maximization, the preferences can be quite different depending on the concrete utility function and its characteristics like risk aversion, reservation level of utility, presence bias, and so forth. On top of that, preferences theoretically need not to be exogenous and stable.

So far we have seen that establishing a preference relation between different contracts, on the one hand, requires the choice of a perspective and, on the other hand, requires picking the criterion for comparison. These choices naturally add lots of extra variety to the vast landscape of potential lending models. Since there are so many different ways of evaluating the same contract, a strict separation between the actual mechanics of a contract and the evaluation of the final outcome is inevitable! To prevent the lending literature from becoming unwieldy, complexity control gets increasingly important and by separating the evaluation module from the contract layer we keep lending theory manageable.

48 We can also imagine combinations of these three *basic* perspectives. A fairness-perspective may combine the lender's and borrower's perspective and judge contracts with a fair distribution of gains between lender and borrower as favourable. A development bank may combine the lender's perspective and the welfare perspective and prefer contracts with high repayment but at the same time low dead-weight costs.

49 A term coined by Herbert A. Simon in 1956 where satisfying one's need suffices. Individuals do not always optimize ([Sim56, p. 129]).

3.6.2 Evaluation criteria in the lending literature

In the literature, most of the time we find borrowers that are modelled as self-interested utility maximizing risk neutral subjects. In particular risk-neutrality may be considered a rather unrealistic assumption.

Q 3.10 The risk attitude of borrowers

What should we realistically assume about the risk attitude of borrowers?

- According to recent neurobiological research uncertainty does not only cause stress, it may in fact be "the essence of stress" [PMF17, p. 164] as the brain uses stress hormones to increase cerebral energy supply in order to reduce uncertainty and avoid surprise. Consequently, mankind should be risk averse.
- On the other hand, obviously there are people that gamble and some even get addicted to it. So the existence of risk-friendly subjects is incontestable.

In summary, the risk attitude of a person seems to be a highly individual attribute and may not even exhibit intra-individual stability. However, perfect risk-neutrality, the tipping point between risk-aversion and risk-seeking, seems to be a rather unrealistic special case.

Concerning lenders, the literature decomposes into two fractions: The bank either just breaks even (T1) or it maximizes its own[50] returns (T2). These two optimization targets are justified by different assumptions about the market structure (T1 or T2) the lender can be faced with (see infobox 3.11).

Q 3.11 Two elementary market structures

Coco presents two optimisation strategies based on different competitive environments:

T1 "Banks' competition is limited to the loan market [...]. The competition process causes the bank to maximize the borrower's welfare, given the cost of fund[s]." [Coc00, p. 201]

T2 "Banks compete contemporaneously both on the market for loans and for deposits [...] the competition process induces the bank to maximize the depositors' welfare with the constraint of keeping the borrowers on their reservation level of utility." [Coc00, p. 201]

Creditors are usually assumed to be risk neutral, too. In contrast to the debtor's risk-neutrality, this assumption is reasoned by the possibility of diversification that sufficiently large banks enjoy. With correlated risks, however, diversification becomes less powerful, thus weakening the argument.

50 the depositor's returns, respectively

Q 3.12 Bank risk taking

There is quite some literature on bank risk taking. The *general* risk attitude of a bank is influenced by a number of different factors like

- monetary policy (see [DNDLV10]),
- market competition (see [BDN05]),
- ownership structure (see [MO14]),
- executive board composition (see [BKS14]).

The risk taking concerning a *single* credit contract then also depends on the bank's possibilities to diversify risks over several non-correlated or negatively correlated engagements.

3.6.3 The term 'first best'

Another topic related to the evaluation of credit contracts is the adjective *first-best*. It is usually used to refer to the outcome that can be achieved in the absence of informational asymmetries. Bolton and Dewatripont speak of first-best if "the agent's choice of action is observable and verifiable" [BD05, p. 130], thus taking as benchmark the absence of hidden actions. But even if there is no hidden information at all, it can still remain unclear which contract is the best (see section 6.3). Thus, the term 'first-best' must be taken with a pinch of salt. It is not at all the ultimately best, it is just the best

- seen from a certain perspective
- under a certain comparison criterion
- that can hypothetically be achieved if there were no hidden information.

As such it can serve as a benchmark.

Q 3.13 The value of trust and honesty

Whether the first-best outcome is a practically useful benchmark in lending theory may be up for discussion. Nonetheless, it can certainly be interesting from a socio economic point of view. If we take the welfare perspective and compare the outcomes with and without hidden information (the first-best outcome) the economic value of trust and honesty is unveiled. The dead-weight costs produced by the informational half of enforcement (p. 24) accord with the economic value of honesty and trust.

Whether universal trust without supervision is a stable equilibrium or invites free-riders to exploitation is a different kettle of fish. Here we are just measuring its theoretical economic value.

Finally, a fair comparison needs fair conditions. For example, if side-contracting amongst villagers is assumed to be costless whilst the bank faces costly auditing (costly state verification) and needs punishments to enforce repayment, then of course joint liability (JL) contracts will be superior to individual liability (IL) contracts from a welfare perspective. But we have to ask: How do villagers enforce their side-contracts, how do they collect their information and set their incentives? In the microfinance literature, the costs of side-contract enforcement sometimes go by the board and the costs of 'social collateral' like the deterioration of social relationships is then neglected. Nevertheless, these costs exist! Karlan observes "direct evidence that relationships deteriorate after default" [Kar07, p.F52]. From the perspective of the lender, however, JL may indeed by superior to IL as he saves enforcement costs by delegating the tasks of information collection and incentive setting to the borrowing group. The message: For a fair comparison of contracts from the welfare perspective it is of vital importance to ensure that all dead-weight costs (losses) that could potentially arise in the course of the contract are indeed considered in the model!

3.7 Contracting

We have learned that credit is a form of reciprocity (see section 1.2) and establishes a connection between two or more legal entities. But how do debtors and creditors get together?

3.7.1 How creditors and debtors technically find each other

There are cases where borrowers and lenders already know each other. For example personal loans when relatives lend each other money, or if an individual uses the internet-banking service of the German Federal Securities Administration (`https://www.bwp-direkt.de`) to buy government bonds, or in case a supplier gives a trade credit to his customer. In these situations creditors and debtors have already found their co-contractor. For all other cases the following means of bringing together creditors and debtors currently coexist: Intermediaries, procuration agencies and bond markets.

The most familiar institutions that serve this purpose are banks. They take deposits and grant loans. And although each client actually has an individual creditor-debtor-relationship with the bank, at heart the bank is an intermediary establishing connections between savers and borrowers.

It is of course also possible to match lenders with borrowers directly and skip the intermediary. Using the power of the internet, peer-to-peer finance platforms establish direct contracts between creditors and debtors.[51] Whereas banks benefit from the interest rate spread between loans and deposits, peer-to-peer platforms take fees to finance themselves. Peer-to-peer lending is not restricted to credit contracts between private individuals. "It allows individual or institutional investors to lend money via its platform to consumers or businesses." [MW13, p. 468] Peer-to-peer lending also does not only refer to one-to-one relationships between creditor and debtor. Many-to-one or one-to-many relations are possible, too. If a credit is provided by many creditors it is also referred to as crowdlending.

Q 3.14 Crowdlending, crowdfunding, crowdinvesting

Concepts similar to crowdlending are crowdfunding and crowdinvesting. Crowdfunding refers to fundraising for businesses or often creative projects. Although it is donation-based, sometimes contributors are promised rewards. If that is the case, we may have a situation that falls under the definition of credit (see section 1.1). Otherwise the compensation promise is missing. In case of crowdinvesting (also called equity crowdfunding) contributors usually become owners of small pieces of the business they invest in.

And finally we have bond markets (credit markets) where creditors and debtors meet. It is divided into the primary market, where participants can issue new debts, and the secondary market, where they can trade bonds already in circulation (see [Ric12, p. 10]). Whilst the primary market de facto establishes new credit connections, the secondary market is just a 'rewiring' of existing creditor-debtor-relationships.

An interesting question is why intermediaries and markets coexist. Why could none of the two mechanisms displace the other over time? One advantage of banks is certainly their expertise and consequently good screening and monitoring abilities. The market, however, benefits from less overhead and lower transaction costs. Financing via financial markets is thus less expensive. A good compilation of models that compare the advantages and drawbacks of market debt and bank debt can be found in [FR08, section 2.5]. The modern formulation of this question, however, might be the question why banks and peer-to-peer finance coexist. The ongoing disintermediation process in the financial sector is driven by progress in IT, by innovative start-ups substituting traditional bank services, and the ever restrictive regulation of traditional banks. In a very well-structured paper Moenninghoff and Wieandt highlight these developments and compare traditional banks with peer-to-peer finance (see [MW13]).

51 Examples for p2p platforms: Lending Club (`http://www.lendingclub.com/`) and Prosper Marketplace Inc. (`http://www.prosper.com/`) in the U.S., or Zopa (`http://www.zopa.com/`) and Funding Circle (`https://www.fundingcircle.com/`) in the European Union.

3.7.2 Informational problems in the negotiation phase

Asymmetric information refers to the fact that one party has more information than its counterpart. What is the relevant information in the contracting phase? It is the borrowers 'type', which can refer to different characteristics like the potential return of his investment project, its riskiness, his own risk-attitude concerning business decisions, and so on. Models usually assume an information asymmetry where the borrower knows more than the lender.[52] The borrower knows his own 'type' whilst the lender cannot see these so-called *hidden characteristics* (see p. 25).

Why is asymmetric information problematic? In the first place, it is just a problem for the lender. He must consider the risk of contracting unawarely with a 'bad' borrower (one that is comparatively unlikely to repay). But the lender can counteract by making the loan more expensive for everyone, to compensate for the increased risk of defaulting 'bad' borrowers. Since the lender cannot distinguish different types of borrowers, we have a *pooling* situation, where all borrowers are treated with the same high interest rate. Meanwhile, asymmetric information is also a problem for 'good' borrowers, as they have to pay the uniform higher interest rate; in fact, 'bad' borrowers get indirectly cross-subsidized by 'good' borrowers.

Siglitz and Weiss show how these higher interest rates can in turn lead to adverse selection: "those who are willing to pay high interest rates may, on average, be worse risks; they are willing to borrow at high interest rates because they perceive their probability of repaying the loan to be low." ([SW81, p. 393]) So increasing interest rates successively leads to drop outs of safer classes of borrowers – the quality of the remaining pool of borrowers gets worse and worse. This is what is called *adverse selection*. The interest rates increase bounces back to the lender and produces new problems.

Adverse selection can lead to quite unaccustomed effects. Usually, in a market with excess demand for credit we would expect the price for credit, i.e. the interest rate, to rise up to a level where demand and supply meet. In markets with adverse selection, however, raising the interest rate worsens the pool of borrowers that is willing to borrow under these conditions. So it may happen, that the expected return to the lender is higher with a lower interest rate (and a qualitatively better pool of borrowers) than with a higher interest rate (but with a pool of borrowers that is rather likely to default). So the law of supply and demand is overturned. The interest rate may remain at a low level, where we have excess demand for credit. This is called *credit rationing* (see [Bal78, p. 172]).

52 This assumption will be scrutinized on p. 80 f.

Q 3.15 Subtypes of credit rationing

Even though the text simply speaks of credit rationing here, one has to be aware of the fact that this term is ambiguous. In the literature, authors have identified several different subtypes of credit rationing.
Baltensperger (1978) already differentiates between "temporary (or "disequilibrium") and permanent (or "equilibrium") rationing" ([Bal78, p. 172]).
Keeton (1979) subdivides equilibrium rationing into two types. One where rationing occurs when a borrower "receives a smaller loan than he would desire at the interest rate charged by the bank" and an alternative type where "the total quantity of loans demanded by the class at those terms exceeds the total quantity supplied by banks" ([Kee79, Preface]), i.e. either a 'haircut' where the size of loans is reduced for all borrowers or a situation where some receive complete funding whilst others go away empty-handed.
Stiglitz and Weiss (1981) even distinguish three different forms of credit rationing:

- "either (a) among loan applicants who appear to be identical some receive a loan and others do not, and the rejected applicants would not receive a loan even if they offered to pay a higher interest rate;" ([SW81, p. 394])
- "or (b) there are identifiable groups of individuals in the population who, with a given supply of credit, are unable to obtain loans at any interest rate, even though with a lager supply of credit, they would." ([SW81, p. 394])
- or another form of rationing where "banks make the provision of credit in later periods contingent on performance in earlier period; banks may then refuse to lend even when these later period projects stochastically dominate earlier projects which are financed." ([SW81, p. 395, footnote 3])

Rationing of type (b), where identifiable groups get no loans at any interest rate, is also called *red-lining* (see [SW81, p. 407]).

Excess demand simply means that some borrowers are denied credit, and as Freixas and Rochet add, even though they are "willing to pay all the price and non-price elements of the loan contract" ([FR08, p. 172]). Thus credit rationing is problematic for the borrowers concerned. But it is critical from a macroeconomic perspective, too: As prospective investment projects may remain unfunded, economic growth and thus the wealth of the whole economy falls short of their actual potential. An example how hidden information can lead to severe global consequences, could be observed in the course of the financial crisis 2007. Certainly not all banks had been affected by the U.S. Subprime Crisis to an extent which brought them to the brink of bankruptcy. But it remained unclear which banks were safe and which were not. Consequently, the risk premium for inter bank loans (measures by the so-called TED spread) increased significantly, the interbank lending market broke down. Despite massive bail-outs, expansive monetary policies and big economic stimulus packages (with even destructive excrescences like the German "Abwrackprämie" in 2009), the world suffered a global economic downturn.

Q 3.16 Historic theories about credit rationing

Let's briefly take up a historic perspective on theories of credit rationing, to see how innovative this informational approach was. Earlier attempts tried to explain rationing

- as a consequence of restrictive monetary policy,
- by sticky interest rates,
- with borrowers whose individual ability to repay deteriorates with credit size.

All these theories turned out to be more or less inconsistent (see [Bal78, p. 170-172]). The nobel price winning economist George A. Akerlof was the first to examine the effects of hidden characteristics on the price the uninformed party is willing to pay in the automobiles market. And he already pointed to the credit market as an instance of hidden characteristics in his famous paper "The market for 'lemons': Quality and uncertainty and the market mechanism" (see [Ake70, p. 497-499]). Meanwhile, the literature on credit rationing is voluminous! The model of Stiglitz and Weiss has been improved (e.g. by [AR09], [Agu09]) and cited numerous times. For details on credit rationing I therefore refer the reader to the original literature. For literature reviews see [Ste10, p. 8-23] or "Microeconomics of banking", chapter 5 ([FR08, p. 171-191]).

To sum it up: If different risk classes of loans can be distinguished, the interest rate – per risk class – can perfectly satisfy two roles (compare p. 34): being the price for capital that balances demand and supply *and* being a risk premium which compensates the lender's risk of defaults. Under asymmetric information, however, we get a precarious compound of both roles. Lenders must charge an average uniform risk premium (risk premium function of interest) which can make capital too expensive for some borrowers (price function of interest) and leads to adverse selection and hence credit rationing.[53]

3.7.3 Possible countermeasures

Information sharing

The most obvious possibility is to mitigate the asymmetric information problem by actively transferring information from the informed party to the uninformed party. Direct communication of the relevant information, however, is impossible (otherwise we would not have an asymmetric information problem). But it may be possible to transfer some information that is at least indicative of the quality of a borrower.

53 For the sake of completeness it should be mentioned that asymmetric information concerning the characteristics of borrowers can also lead to overinvestment, instead of credit rationing. See [DMW87].

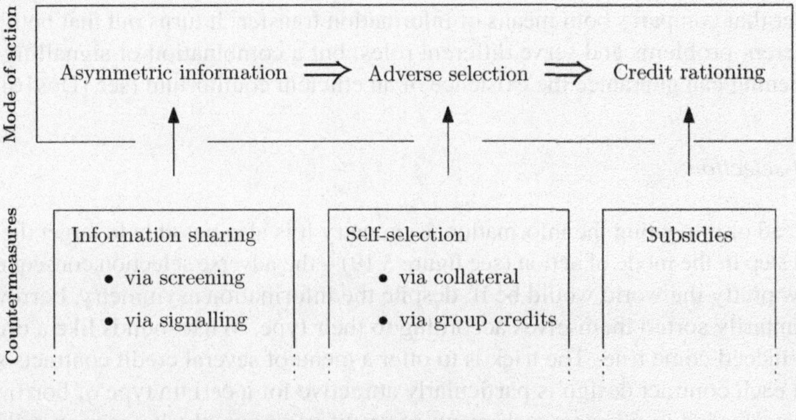

3.19 Informational problems in the connection phase and solutions
This graphic summarizes how informational problems in the contracting phase can have adverse effects and which countermeasures can mitigate the problem at each stage.

In models where the informed party has the leading role in transferring indicative information this information transfer is called *signalling*. The informed party sends some signal about its quality. This signal can be the college degree of a job applicant indicating his quality as a worker, a limited warranty provided by the seller of a 'good' car, the business plan of a start-up indicating the quality of their business idea, the rating of a credit rating agency (paid by the borrower), or even such simple information as the maximum interest rate a borrower is willing to accept[54]. In fact, all information that is positively correlated with the quality of the borrower can serve as a signal.

The complementary version of an active transfer of indicative information is *screening*. Here the uninformed party takes the lead and collects indicative information about potential contractors. Practically this is realized via credit bureaus like the Schufa Holding AG (GER), Experian or Equifax (UK), or TransUnion (US).

Characteristic for signalling and screening is that indicative information is actively sent (by the borrower) or collected (by the lender) to alleviate the information asymmetry at its very origin. A legitimate question is whether both are interchangeable. It turns out that there are some fundamental differences in their modes of operation.

54 Kawai, Onishi and Uetake use data of an online peer-to-peer loan market (namely Prosper.com) and show that the reserve interest rate a borrower publicly posts signals his creditworthiness (see [KOU14, p. 2]).

In order to stick to the central theme of this thesis, I refer the reader to a very recent paper that compares both means of information transfer. It turns out that both face different problems and serve different roles, but a combination of signalling and screening can guarantee the existence of an efficient equilibrium (see [Dos16]).

Self-selection

Instead of mitigating the information asymmetry it is also possible to target the second step in the mode of action (see figure 3.19) – the adverse selection consequence. How pretty the world would be if, despite the information asymmetry, borrowers voluntarily sorted themselves according to their type. What sounds like a dream, can indeed come true. The trick is to offer a menu of several credit contracts such that each contract design is particularly attractive for a certain type of borrowers. If we succeed in offering such menu of credit contracts, the borrowers will sort themselves according to their type, which we call *self-selection*.

Credit contracts exhibit a variety of parameters (see section 3.2). Melnik and Plaut say that credit contracts "are most properly seen as 'packages' of loan terms including the interest rate, the commitment fee, maturity, the credit quota or takedown limit, and collateral arrangements ... from which borrowers may select those that appeal most to them." [MP86, p. 434, in the context of loan commitment contracts] All these loan terms are potential candidates to induce self-selection. In the literature we find self-selection induced by differing interest rates and collateral requirements (see [Bes85]) but also along different degrees of joint liability (see [Gha00]) or even different loan sizes (see [MR88]).

Let's throw a quick glance at the technical details of self-selection. Bester, in his paper "Screening vs. Rationing in Credit Markets with Imperfect Information" ([Bes85]), shows how self-selection can be induced with collateral. He starts off with two types of borrowers, one with a risky investment project and another one with a safe project (whilst mean project returns are identical), which are indistinguishable for the bank. Bester analyses the slope of the interest-collateral-indifference-curves of each type and shows that the safe type is better off with a high collateral contract and the risky type chooses a low collateral contract, respectively (see marginal rate of substitution between collateral and interest in [Bes85, p. 853])[55]. The bank then offers two contracts, one contract with a high interest rate and low collateral require-

55 Notice, that an additional layer of complexity can be introduced when, not only the success of the investment project, but also the value of collateral is considered to be stochastic. As shown by [Nii09], this can flip the result of [Bes85] upside down.

ments and another one with a low rate and high collateral requests. Figure 3.20 illustrates the principle.[56]

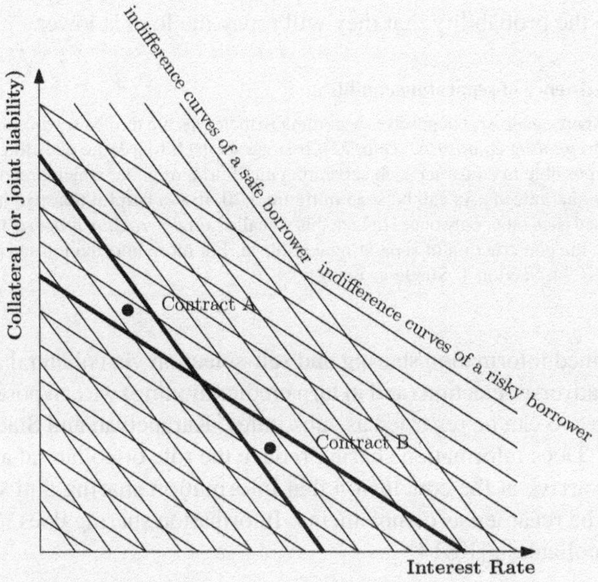

3.20 Self-selection principle
This chart illustrates how different contracts can selectively attract different types of borrowers. Each point of the plane represents a credit contract with a certain interest rate and a certain amount of collateral (or joint liability). For save borrowers contract A is more attractive than contract B, for risky borrowers contract B is more attractive than contract A.

Another example is Ghatak, who presents a self-selection mechanism for a group lending environment in "Screening by the company you keep" ([Gha00]). There, groups of borrowers of safe type choose the contract with low interest and high joint liability whereas groups of risky borrowers choose the contract with high interest and low joint liability (see Proposition 2 in [Gha00, p. 615]).[57]

56 Models about self-selection may include additional features like iso-profit lines for the lender, or optimal contract menus under competition (see figure 2 in [Bes85, p. 853]). And, depending on the probability distribution of the investment project, the indifference curves may not be linear.

57 In his paper, group formation is an endogenous process. In lemma 1, however, Ghatak shows that borrowers will always form groups of homogeneous risk classes in his model. Also see p. 79.

Intuitively, these self-selection mechanisms exploit that a borrower with a safe project has a comparatively low aversion against collateral (or joint liability) because his type makes it unlikely that he will be affected much by these parts of the contract. Simultaneously, borrowers with a risky project are willing to pay higher interest rates because the probability that they will repay the loan is lower.

Q 3.17 Existence of separating equilibria

Equilibria where agents sort themselves according to their type are termed *separating equilibria* in contrast to *pooling equilibria* (see p. 72). It suggests itself to ask the question of whether it is always possible to construct such separating equilibria, or under which preconditions its existence is guaranteed. As can be seen in figure 3.20, it was helpful that two indifference curves crossed each other only once. In fact, this so-called *single-crossing property* is a standard argument in the construction of separating equilibria. For more information see [MCWG95, p. 453] or [KS94, Section 4. Single-crossing].

Having explained information sharing and self-selection via collateral as two means of mitigating adverse selection (and in turn credit rationing) we can pose the question whether these two can be regarded as substitutes. Karapetyan and Stacescu wrote a paper asking "Does information sharing reduce the role of collateral as a screening device?" and arrive at the conclusion that information sharing and self-selection should rather be regarded as complements. Information sharing does not reduce the incidence of collateral. [KS14]

Is there a difference between information sharing and self-selection, at all? Notice that through the borrowers' contract choice we implicitly gain some information, as by their choice the borrowers reveal their type. This self-selection process is therefore also called "market screening". And, in addition, Kreps and Sobel point out that the choice of a contract can be seen as a signal (see [KS94, p. 862]), so it could even be called signalling. However, one case exhibits some explicit information transfer (screening and signalling) and then the lender decides whom he offers which contract. In the other case we have no explicit ex ante information transfer, the borrower chooses a contract (self-selection) and only afterwards we get some implicit information transfer as a by-product. So we have two distinct situations:

- (explicit) information transfer $\rightarrow$ contract choice by the lender
- contract choice by the borrower $\rightarrow$ (implicit) information transfer.

Sometimes they even lead to different results. With information sharing risky borrowers end up with credit contracts characterized by rather high amounts of collateral. Whereas under self-selection risky borrowers have the contracts with little collateral (see [Bes85]).[58] So screening/signalling and self-selection are in fact different concepts, and the term "market screening" just blurs the line.

Subsidies

Finally, we have to look at a third possibility to parry credit rationing. If some individuals are not granted credit by the usual route, why not simply supply the rest with credit via subsidies? In fact, state-owned development banks tried this solution, but Armendariz and Morduch explain that subsidized credit exhibit several severe drawbacks (see section 1.3 in [AM10, p. 9]): Other lenders where pushed out of the marked, the interest rate lost its role as a rationing mechanism that filters high quality investment projects and instead credit was often allocated "on the basis of politics or social concerns" ([AM10, p. 11]), the bank's incentive to collect savings deposits declined as well as the incentive to insist on repayment. [HCG06] find that "the presence of subsidies is … associated with higher MFI costs". Thus, in theory subsidies sound like a simple solution to make up for the consequences of credit rationing, but in practice they often fail. For a comprehensive discussion of the topic and suggestions to improve the efficacy of subsidies, see [AM10, Chapter 10].

3.7.4 Special case: Group formation in group lending

So far we have discussed how lenders and borrowers come together and connect with each other. In group lending situations an additional contracting topic is the formation of borrower groups. Regarding group formation we encounter a quality and a quantity issue.

Quality: The types of group members

Quality refers to the question which types of borrowers are bond together in one group. The theoretic research about endogenous group formation has probably begun with Ghatak's paper "Group lending, local information and peer selection" ([Gha99]). Under various degrees of freedom (e.g. many types of borrowers, unbalanced proportion of types, heterogeneity in wealth levels of borrowers, group size exceeding two) he shows that self-selection of group members leads to positive assortative matching, i.e. safe borrowers join up with safe borrowers and vice versa. Even if transfer payments between borrowers are allowed and a risky borrower could compensate the safe borrower for the increased risk, borrowers keep forming

58 Empirical evidence suggests that high amounts of collateral are associated with high risk, see [BU90].

uniform groups because the compensation risky borrowers are willing to pay is less than what safe borrowers require (see [Gha00, p. 609, 610]).

However, there is a also literature showing that this positive assortative matching result is not as general as thought. For example Guttman (2008) shows that positive assortative matching does not necessarily prevail under the threat of not being refinanced in case of group default ([Gut08]) – a common enforcement method in microfinance. Banerjee and Sethi (2008) analyse different forms of intra-group transfers. Their result: Certain transfer schemes may lead to heterogeneous group formation ([BS08]). Or Katzur and Lensink (2012), which spice up a simple group lending model with correlated project returns, end up with the finding that in some cases "the positive assortative matching result may break down" ([KL12, Section 4.3]). [GL14] and their analysis of asymmetric group loans also features heterogeneous matching.

The literature about endogenous group formation asks how borrowers will from their joint liability groups. But it is critical to uncover the implicit assumptions these models make.

First of all, do borrowers have much choice, i.e. is the pool large enough? In real life, borrowers face the challenge of finding a partner who they know and trust, who needs a loan at the same time, of roughly the same size (to make mutual insurance possible). If the borrower is lucky and finds such a person, he may not be left with much variety concerning the type of his partner. So it remains doubtful if a borrower really has much choice.

Second, can borrowers choose sensibly, i.e. are they skilled enough? Given there is sufficient choice, a borrower would have to assess the risk profile of a bunch of other borrowers in order to adequately choose his partner. There may be inherent project risks (e.g. associated with the production technology), as well as market risks (e.g. a drop in demand for the produce) or environmental risks (e.g. weather conditions). Admittedly, the borrower may have some additional information about his peers compared to the bank (asymmetric information assumption), but the bank may know much more about market and environmental risks. Moreover, information is not enough. It also takes the skill to process this knowledge. Aggregating several risks to a single risk figure poses a severe challenge – even to sophisticated and highly skilled risk departments of banks and insurance companies. So it remains doubtful whether a borrower can judge a potential partner's risk precisely enough.

Of course you could argue that, taking the whole world into account, the pool will be large enough, and if the borrower gathers enough knowledge, he will be able to conduct proper risk analyses; so endogenous group formation should work. With this exaggerated formulation, however, it becomes apparent what the models mentioned above omit: It is the cost of group formation! As Chatterjee and Sarangi (2005) mention "it may be easier to form groups with some people than others, regardless of the risk characteristics of these individuals. This could be for reasons like the person's proximity in terms of the work place, language, religion, ethnicity, or just plain neighbourhood." ([CS05, p. 2]) So the costs of group formation may spoil the idea of effective endogenous group formation.

Against that background it seems reasonable to build group lending models without (positive or negative) assortative matching. One example for such a model is [AG00], where joint liability groups are formed randomly. They find that randomly formed joint liability groups can still alleviate the problem of credit rationing. Another example of random group formation is [BZ06]. They find that only safe borrowers will accept joint liability. In fact all joint liability results should be tested for their robustness under random group formation.[59]

So far, the decisive characteristic has been the risk of a borrower's investment project. Most of the respective lending literature focuses on sorting along risk classes – maybe because it is the most relevant characteristic in the context of credit markets, especially with regard to charging risk premiums (see p. 34). But of course we could consider other characteristics of a borrower, for example his risk attitude, and ask the question again whether borrowers sort themselves homogeneously with respect to this new characteristic.[60]

Quantity: The number of group members

The quantity issue of group formation refers to the optimal size of a joint liability group. Examples for papers examining the aspect of optimal group size are [BS12] and [Ahl15].

To optimize group size we need to balance favourable and detrimental effects of an increasing number of group members. For example risk diversification is better in bigger groups. Concerning social pressure larger groups may be better because of the sheer number of peers. But they could also perform worse due to increased anonymity amongst group members and free-riding may become easier. In general, social ties between group members play an important role. If they are too weak,

59 An even more radical assumption would be borrowers that do not even know their own type! The justification of this assumption is simple: it is the same argument that finding out a borrower's type can be very costly, but applied to the borrower himself. Of course the assumption that a borrower can only guess his own type would not only affect group formation but all aspects of contracting.

60 Schulhofer-Wohl, for example, wrote a paper called "Negative assortative matching of risk-averse agents with transferable expected utility" [SW06]. It refers to the marriage market but nevertheless it is quite general and presumably applicable to credit markets, too.

intra-group monitoring (see section 3.4.4) and risk-sharing may suffer. If they are too strong the lender may be faced with collusion. Furthermore, the individuals' sensitivities to social sanctions are crucial (see [Imp98]).

Concerning optimal group size, the driving forces may be known, but theoretically deriving a precise optimal number of members will most likely fail. It is difficult to imagine how soft factors like social ties between people can be measured and formalised accurately enough. In reality feelings like sympathy or antipathy between individual group members may change the whole game. However, the theory of optimal group size may be able to illustrate some general principles.

3.8 Further topics related to credit contract theory

Basically, credit contracts are a tool. If we, for example, consider a knife, we can ask questions concerning its construction like what material the blade is made of, how long it is, what its balance is, how the blade can be attached to the handle, etc. Analogously, in this thesis the design and construction of the 'tool' credit is discussed.

But, beyond credit contract theory – or what [FR08] call the lender-borrower relationship – lies a whole world of further topics that are associated to credit. Banking theory for example (see [FR08]) includes a variety of other issues like transformation problems within a lending institution. Let's throw a quick glance at some of the topics in the vicinity of credit contract theory:

Systemic topics: financial intermediation theory (e.g. transformation problems, cost advantages over markets); systemic risks and the capitalisation of banks; ...

Macroeconomic implications: on liquidity and monetary policy; on economic growth, stability and crises; ...

Social consequences: the empowerment or harassment of women by microcredit; effects on the gap between the rich and the poor; ...

Moral questions: excessive interest, fair interest and Islamic Banking; inter-generation fairness of public debt; ...

The purpose of this list of examples is to sensitize the reader that many problems associated with credit are not problems of the 'tool' credit but rather an issue of its use – or misuse. Staying in our analogy: Building a functioning knife is one thing; using a knife is a whole new ball game.

4 Simplicity: The interplay of product layer modules

Whilst monolithic models combine a whole chain of interacting modules, the modular research strategy focuses on individual modules and their improvement by abstracting the rest of the system to simple input and output signals. And as Nisan and Schocken emphasise in the context of computer science: "…a good modular design implies just that: You can work on the individual modules independently, while completely ignoring the rest of the system." [NS05, p. 3]

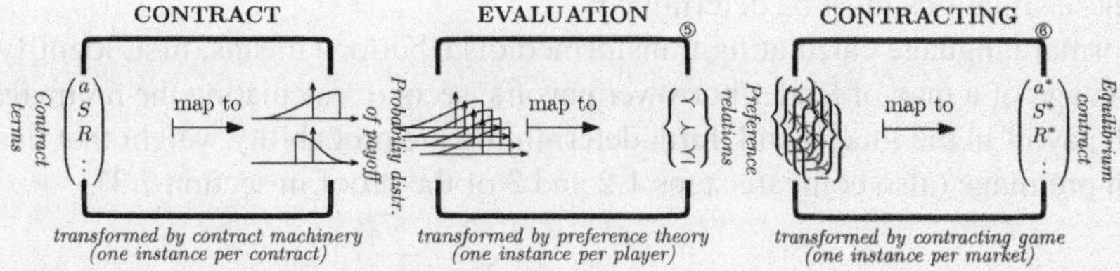

4.1 Combining product layer modules
Interfaces between product layer modules. (Large version on p. 99.)

C. Prem, *The Theory of Credit Contracts*,
https://doi.org/10.1007/978-3-658-29362-8_4

For readers unaccustomed to a modular research strategy, non-monolithic models may feel unfinished. This feeling may arise in chapters 6 and 7 where we study the consequences of different reciprocity agreements – without caring e.g. about their enforcement. The impression may also arise in chapters 9, and 10, 11, 12, which concentrate on the improvement of payback-enforcement schemes – without talking e.g. about credit markets.

To cure the feeling of incompleteness of individual modules, this chapter will

1. demonstrate how the product layer modules can be assembled to 'complete' models for *concrete applications*
2. explain why it makes lots of sense to not perform this assembly within the scope of *basic research*.

Note that this chapter contains references to future chapters where the details to some of the examples will be elaborated. So at first reading, the reader should not worry about the details too much and try to understand the principles instead. At the end he may wish to come back and reread this chapter.

4.1 Contract: From loan terms to payoff distributions

As explained in section 2.3 we can abstract from all the technical details of a credit contract and regard it as a probability distribution of payoffs. The distribution of payoffs tells how likely it is that a player gets a certain payoff at the end of the credit relationship.

In this first section we will study examples on how to calculate this distribution of payoffs from the loan terms of a credit contract. The fundamental principle is always the same: First, the possible payoffs of a player must be identified. Second, the situations that lead to a certain payoff must be investigated. Third, the probability of these situations must be determined.
In formal language calculating transformed distributions means, first, identifying the image of a map of lender/borrower payoffs, second, calculating the preimage of each payoff in the image, and third, determining the probability weight that lies on such preimage (also compare steps 1,2 and 3 of the proof in section 7.3).

4.1.1 Examples for calculating the distribution of borrower payoffs

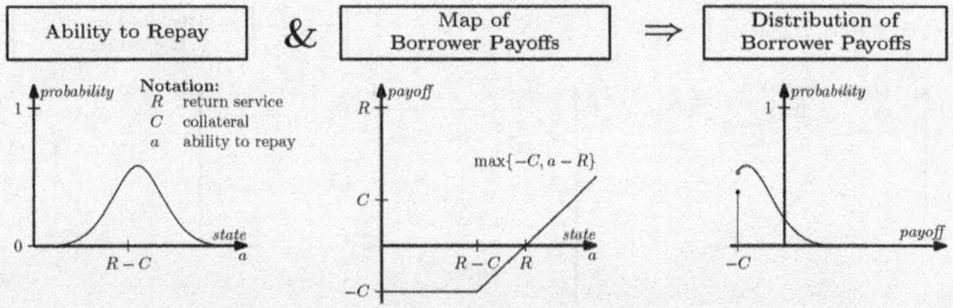

4.2 Calculating the distribution of borrower payoffs (SDC)
Distribution of borrower payoffs in a standard debt contract (see chapter 7, Lemma 1).

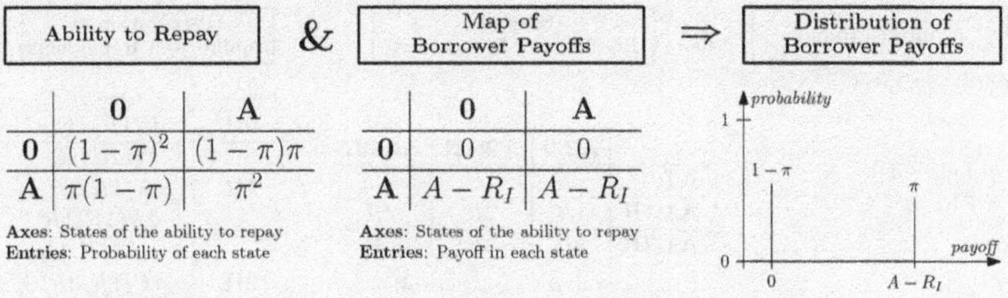

4.3 Calculating the distribution of borrower payoffs (discrete IL)
Distribution of borrower payoffs in a discrete IL contract (see section 6.2).

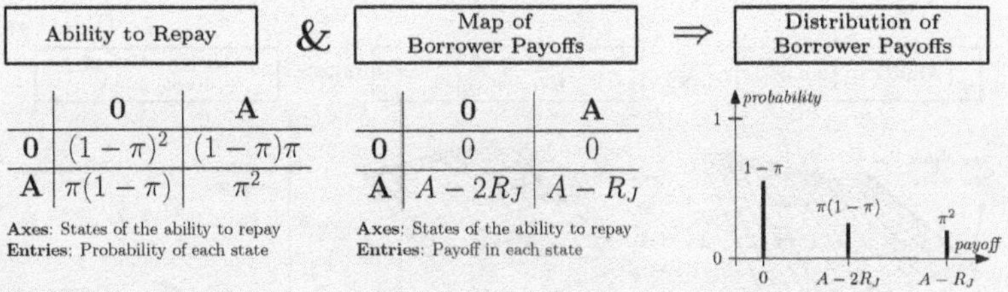

4.4 Calculating the distribution of borrower payoffs (discrete JL)
Distribution of borrower payoffs in a discrete JL contract (see section 6.2).

4.1.2 Examples for calculating the distribution of lender payoffs

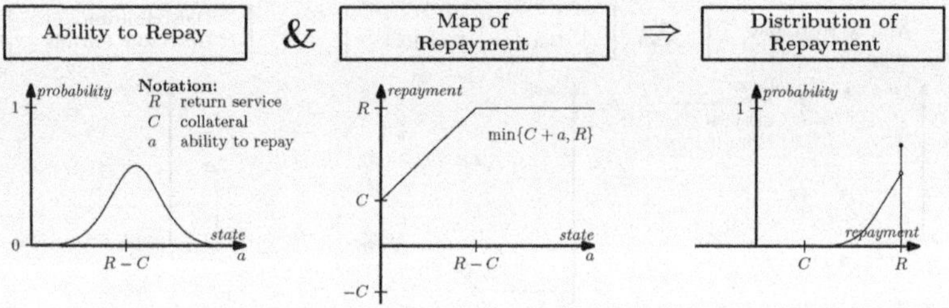

4.5 Calculating the distribution of repayment (SDC)
Distribution of repayment for a standard debt contract (see chapter 7).

Ability to Repay & Map of Equilibrium Repayment ⇒ Distribution of Equilibrium Repayment

Joint CDF
$F(a_1, a_2) \in [0, 1]$

	A2:0	A2:1R	A2:2R
A1:0	0	$1R$	$2R$
A1:1R	$1R$	$2R$	$2R$
A1:2R	$2R$	$2R$	$2R$

Axes: States of the ability to repay
Entries: Repayment in each state

repayment	*probability*
0R	$F(1R, 1R)$
1R	$F(2R, 1R) + F(1R, 2R) - 2R(1R, 1R)$
2R	$1 - F(2R, 1R) - F(1R, 2R) + F(1R, 1R)$

4.6 Calculating the distribution of repayment (discrete map)
This example shows how to calculate distributions with the maps of appendix B.

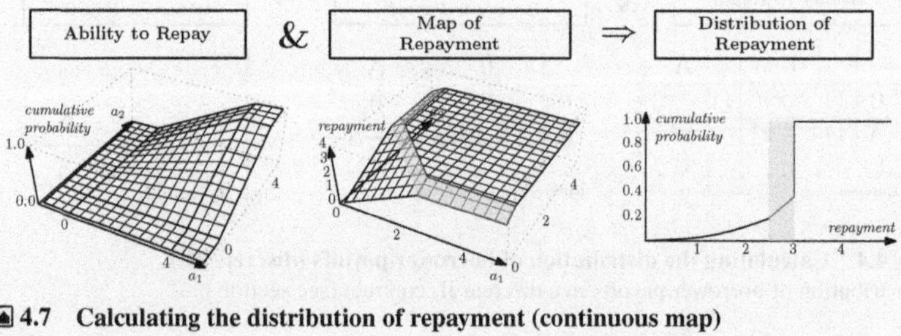

4.7 Calculating the distribution of repayment (continuous map)
This example shows how to calculate distributions for a continuous map of repayment.

Q 4.1 Calculating transformed distributions with Mathematica

When calculating the example in figure 4.7 by hand we need to deal with numerous distinctions of cases. This tedious task can be handed over to a computer and the example can be reproduced by executing the following short Mathematica-Code.

```
"ABILITY_TO_REPAY_(CDF)"
distr =UniformDistribution[{{0,4},{0,4}}];
Plot3D[CDF[distr,{a1,a2}],{a1,-1,6},{a2,-1,6}]

"REPAYMENT_FUNCTION"
r=1.5;  c=0;  l=1;
repayfunc[a1_,a2_]:=Piecewise[{
{c+a1,a1<=r-c},
{r,a1>r-c && a2>=r-c},
{r+Min[r-(c+a2),(c+a1)-r,l],a1>r-c &&a2<r-c}}];

"MAP_OF_REPAYMENT"
repaymap[a1_,a2_]:=repayfunc[a1,a2]+repayfunc[a2,a1];
Plot3D[repaymap[a1,a2],{a1,-1,6},{a2,-1,6}]

"TRANSFORMED_DISTRIBUTION_(CDF)"
repaydistr =TransformedDistribution [repaymap[a1,a2],{a1,a2}\[Distributed] distr ];
Plot[CDF[repaydistr,x],{x,-1,6},AxesLabel->{"repayment","cumulative_probability"},PlotLabels->"CDF"]
```

Let's supplement the last three examples with some more explanations. If you want to incorporate loan size, shift the repayment function down by the size of the loan. This results in a left-shift of the distribution of lender payoffs compared to the distribution of repayment.[61]

The example in figure 4.6 illustrates how it is often convenient to work with joint cumulative distribution functions instead of working with marginal distributions and copulas separately or integrating over (not always existing) probability density functions. This is especially the case when continuous probability distribution are combined with discrete maps with its rectangular regions. The map of repayment in figure 4.6 is that of B.3 or B.8 or B.12. Thus, with calculating the distribution of repayment for a single map we cover a whole group of repayment games – which is an example how a modular research strategy not only provides structural clarity and complexity control but also saves effort.

Figure 4.7 provides a sample calculation for a continuous two-dimensional map. We recycle the repayment function with joint liability and collateral from figure 3.8. If you take one repayment function for borrower 1 and one (a symmetric version) for borrower 2, then the sum of both is the repayment received by the lender. A short Mathematica code can save your time for the tedious calculations necessary to establish the transformed distribution, see box 4.1.

Note that all these calculations can also be done using a map of deadweight loss, but a third subsection with examples for "Calculating the distribution of deadweight loss" is omitted. It is always the same boring procedure.

61 Only relevant if loans of different size are to be compared.

4.1.3 The inside and the outside of the contract module

Inside

The examples above are all different: Different abilities to repay, different probability distributions, different perspectives (borrower or lender), different maps of equilibrium payoffs, discrete and continuous models, graphical illustrations and formal presentations. However, it is of vital important to grasp the ever repeating common pattern behind this variety of different examples. The reader should make sure that he can match the steps of the general principle (described in the beginning of this section) with each individual example.

Note how all the information about marginal distributions, dependence structures and the results of potential purpose-enforcement modules nicely cumulate in the joint CDF of the ability to repay. And note how all the details of punishment strategies and payback-enforcement are condensed in the map of equilibrium repayment, map of borrower payoff and map of deadweight loss. Both information then come together and form the distributions of payoffs which are characteristic for these loan terms, see figure 4.8.

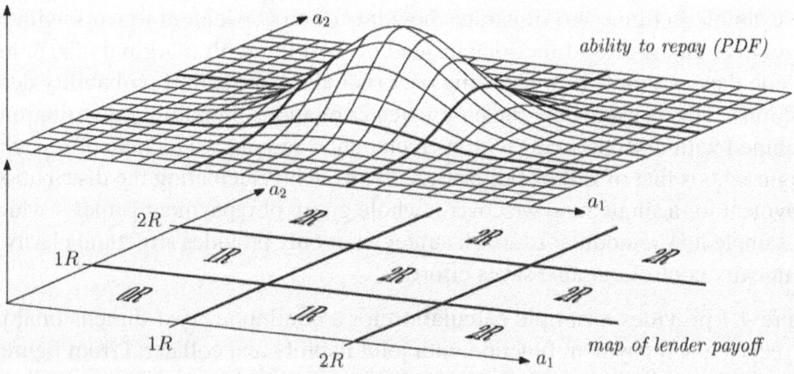

4.8 Combining ability to repay and map of repayment
This figure illustrates how an ability to repay and a map of repayment match together. The map tells the payoff in each state. The ability to repay tells the probability of each state.

Outside

In this section we have studied several examples of how to calculate the distribution of payoffs for each player – which is finally the interface that connects the contract modules to the evaluation module. It also constitutes the border between credit contract specific topics and topics that are not credit contract specific, because whether the distribution of payoffs is borne by a credit contract or e.g. an insurance contract is irrelevant for the following modules. Figure 4.9 illustrates the interfaces of the contract module.

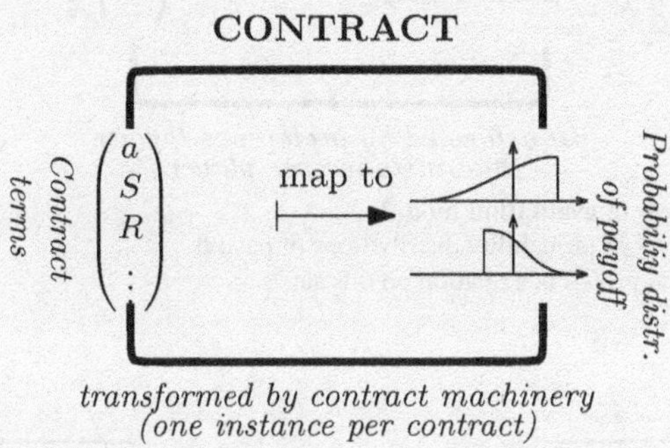

4.9 Interfaces of the contract module
DATA-INPUT: Contract terms like service S, return service R, ability to repay a, etc.
DATA-OUTPUT: Probability distribution of payoffs, one distribution for each party.

4.2 Evaluation: From payoff distributions to preference relations

The evaluation module turns a menu of payoff distributions into a preference relation. Of course, the distributions represent contracts, but it is not necessary to know that, in order to calculate the preference relation. Accordingly, the symbols that describe the distributions in the examples below have some meaning in the context of credit contracts, but in the evaluation module we can forget about its meaning. It is important to understand that intellectual abstraction from concrete contracts! In the evaluation module, all the information about the origin of the set of distributions is not important any more.

The formal tools to perform the transformation in the evaluation module are provided by preference theory (e.g. see first and second order stochastic dominance and mean preserving spreads in [DD05, Chapter 4]). To understand the examples below it is enough to know how to calculate the expected value (for risk-neutral players) and how to compare two CDFs using the integral conditions for mean preserving spreads (for risk-averse and risk-friendly borrowers).

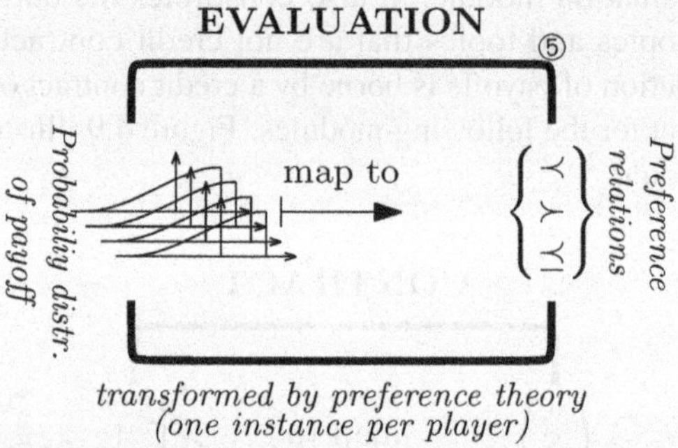

4.10 Interfaces of evaluation module
DATA-INPUT: A set of probability distributions of payoffs
DATA-OUTPUT: A preference relation on this set

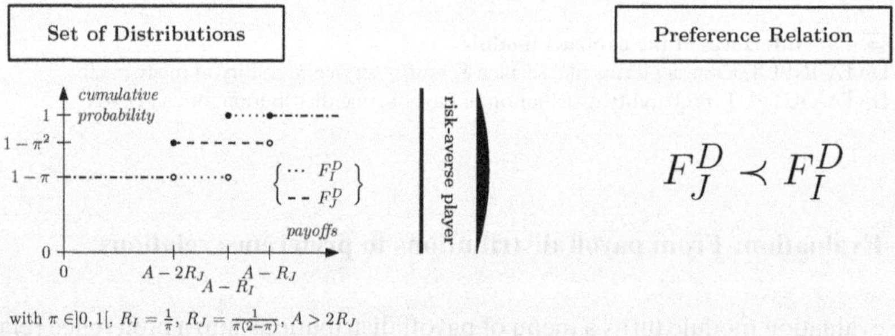

4.11 Evaluating sets of distributions (two discrete distributions)
Evaluation of two discrete distributions by a risk-averse player (see chapter 6).

It is easy to derive some more examples from figure 4.11. If we take as evaluation criterion risk-neutrality, then the preference relation becomes $F_J^D \sim F_I^D$, and evaluation by a risk-friendly player turns the relation upside down. (For the proof, see chapter 6.) Another example would be to let $R_I < A < 2R_J$ and assume a player with loss aversion. This player would choose $F_J^D \prec F_I^D$ since F_J^D had a negative realisation with positive probability.

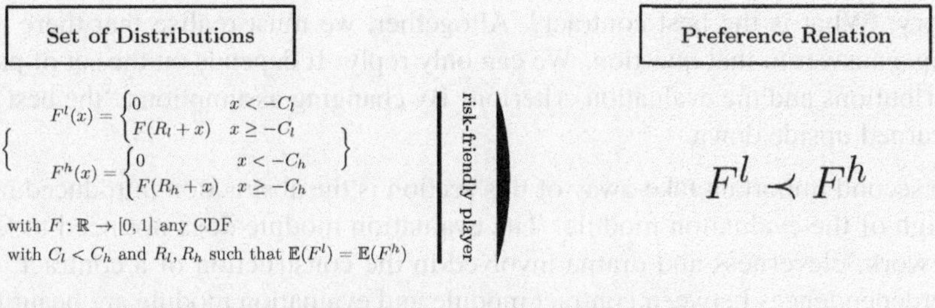

4.12 Evaluating sets of distributions (two mixed distributions)
Evaluation of two mixed distributions by a risk-friendly player (see chapter 7).

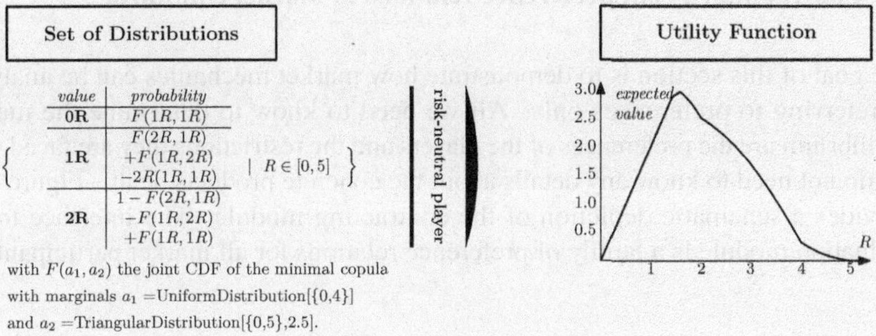

4.13 Evaluating sets of distributions (infinite set of distributions)
This is an example with a set of infinitely many payoff distributions. Distributions are taken from figure 4.6. A risk-neutral player would calculate the expected value of each distribution and the preference relation can conveniently be represented by a utility function, i.e. $x \preceq y :\Leftrightarrow \mathbb{E}(x) \leq \mathbb{E}(y)$.

Q 4.2 Calculating utility functions with Mathematica

Calculating the algebraic term of the utility function in figure 4.13 by hand involves dealing with lots of distinctions of cases. As the kinks of the function plot indicate, the result is a piecewise defined function. The calculations can be reproduced by applying the following short Mathematica-Code.

```
"DEFINING_A_FIXED_F(a1,a2)"
a1=UniformDistribution[{0,4}]; (* ability to repay of borrower 1 *)
a2= TriangularDistribution [{0,5}]; (* ability to repay of borrower 2 *)
copula="Minimal"; (* dependence structure *)

distr = CopulaDistribution [copula,{a1, a2}];(* joint probability distribution *)
F[a1_,a2_]:=CDF[distr,{a1,a2}]; (* joint cdf *)

"EXPECTED_VALUE"
expRepay[R_]:=(2-F[2R,1R]-F[1R,2R])*R; (* expected repayment as a function of R *)
Plot[expRepay[R],{R,0,6}] (* illustration of the function expRepay[R] *)
```

Now, in a set of distributions, what is 'the best' distribution? Translated to lending theory: What is the best contract? Altogether, we must realise that there is no unique answer to that question. We can only reply: It depends on the set of payoff distributions and the evaluation criterion! By changing assumptions, 'the best' can be turned upside down.

The second important take-away of this section is the abstraction introduced in the design of the evaluation module. The evaluation module does not need to know the work, cleverness and drama involved in the construction of a contract. The interdependences between contract module and evaluation module are beautifully captured by the payoff distribution interface.

4.3 Contracting: From preference relations to market equilibria

The goal of this section is to demonstrate how market mechanics can be analysed by referring to preferences only. All we need to know to determine the market equilibrium are the preferences of the players and the restrictions they are faced with. We do not need to know any details about the concrete products dealt. Figure 4.14 provides a schematic depiction of the contracting module. The interface to the evaluation module is a family of preference relations for all market participants.

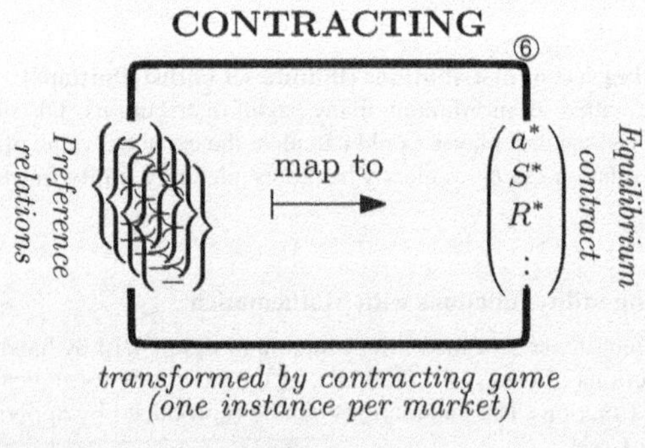

4.14 Interfaces of contracting module
DATA-INPUT: for each player a preference relations on the set of contracts
DATA-OUTPUT: market equilibrium contract(s)

The tools that are needed to calculate the interior of the module are provided by game theory. A contracting game is a game theoretic game, most often a sequential game, in which a provider makes an offer and a customer chooses between taking or leaving it. In case the provider offers several options, the customer may also choose between different types of products. The deals that are actually closed, can be determined by an equilibrium analysis.

The examples that follow present the simplest setting in which we can analyse market mechanics. The main information we are working with are preferences. However, it only gets interesting if we add additional assumptions about the information distribution (hidden characteristics), assumptions about the relative attractiveness of possible outside options (OO), and assumptions about the market structure (T1 or T2, see infobox in subsection 3.6.2). The various possible deals are denoted by P with various indices to distinguish them. Notice that not every type has access to every deal.[62] Since we consider a sequential game, we have to decide who moves first. The market equilibrium is then determined by performing a backwards induction, i.e. the *offering party* determines its action considering the reaction of the *choosing players*.

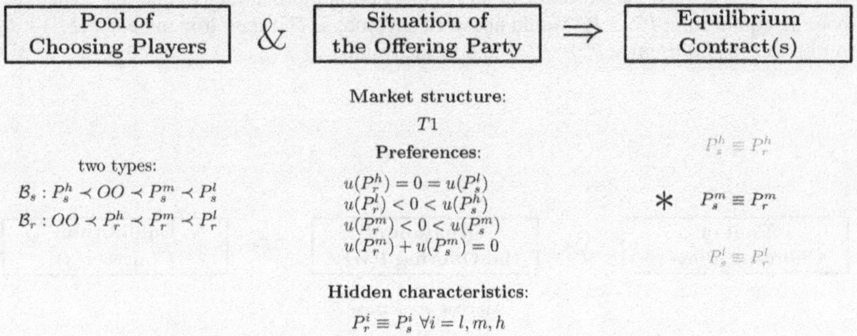

4.15 Determining the market equilibrium (pooling)
In our first example the first-mover suffers from hidden characteristics, i.e. not all products are clearly distinguishable. Thus it can only provide combinations of (indistinguishable) contracts. It has to exactly break even (T1). Since the combination $P_s^l \equiv P_r^l$ leads to a negative utility it bows out. The combination $P_s^h \equiv P_r^h$ triggers a drop out of the $\mathcal{B}_s$-type but then delivers zero utility. However, the combination $P_s^m \equiv P_r^m$ keeps the $\mathcal{B}_s$-type in the market and also achieves break-even, whilst providing a higher sales volume.

Let's consider a few variations of the example in figure 4.15. As we have just seen, there is no combination that gives the uninformed party a positive utility. Thus, a similar model but with market structure T2 instead of T1 results in the same

62 Think of it as a car owner in the market for lemons: It can only sell his own car, although at different prices. Or think about it as a borrower: he has only access to his own ability to repay but can close contracts with various degrees of joint liability.

market equilibrium. However, switching from T1 to T2 would indeed change the equilibrium to the $P_s^h \equiv P_r^h$ combination, if $u(P_s^h)$ were positive.

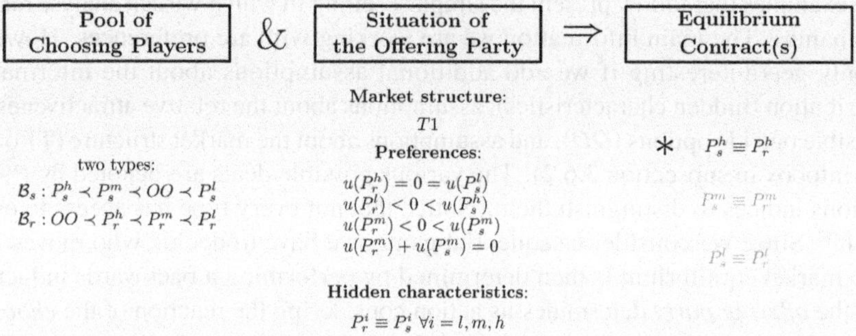

4.16 Determining the market equilibrium (rationing)
Imagine we assumed the existence of an outside option more attractive than P_s^m. Then the pooling equilibrium $P_s^m \equiv P_r^m$ would not work anymore as $\mathcal{B}_s$ types drop out and $u(P_r^m) < 0$. A rationing equilibrium at $P_s^h \equiv P_r^h$ would be the result.

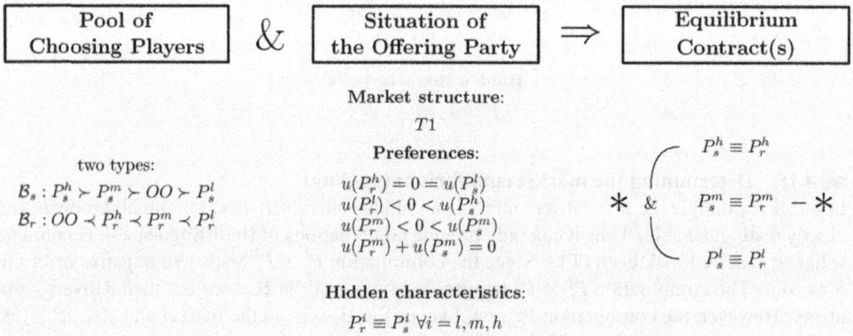

4.17 Determining the market equilibrium (pooling & self-selection)
So far, both types exhibited the same preference ordering with respect to $P_s^i \equiv P_r^i$ combinations. But there are markets where this is not the case. (E.g. borrowers of differing risk-attitude prefer different interest-to-securities-ratios, see chapters 6 and 7.) The following market model offers a pooling equilibrium. In addition it offers a self-selection equilibrium where the choosing players sort themselves automatically. (For more information on markets with self-selection equilibria see subsection 3.7.3.)

Pool of Choosing Players	&	Situation of the Offering Party	⇒	Equilibrium Contract(s)

Market structure:

$T1$

Preferences:

two types:

$\mathcal{B}_s : P_s^h \prec OO \prec P_s^m \prec P_s^l$

$\mathcal{B}_r : OO \prec P_r^h \prec P_r^m \prec P_r^l$

$u(P_r^h) = 0 = u(P_s^l)$
$u(P_r^l) < 0 < u(P_s^h)$
$u(P_r^m) < 0 < u(P_s^m)$
$u(P_r^m) + u(P_s^m) = 0$

* $P_r^h \& P_s^l$
* $P_s^m \& P_r^m$

Hidden characteristics:
none

4.18 Determining the market equilibrium (with break-even)
Assume we could get rid of the hidden characteristics assumption by applying screening or signalling techniques. (For more information on screening and signalling see subsection 3.7.3.) What would be the market equilibrium in this case? We have multiple equilibra, either the pooling menu $P_s^m \& P_r^m$ or the separating menu $P_s^l \& P_r^h$. The first equilibrium works because the slightly too low utility of P_r^m is exactly compensated by the slightly too high utility P_s^m.

Pool of Choosing Players	&	Situation of the Offering Party	⇒	Equilibrium Contract(s)

Market structure:

$T2$

Preferences:

two types:

$\mathcal{B}_s : P_s^h \prec OO \prec P_s^m \prec P_s^l$

$\mathcal{B}_r : OO \prec P_r^h \prec P_r^m \prec P_r^l$

$u(P_r^h) = 0 = u(P_s^l)$
$u(P_r^l) < 0 < u(P_s^h)$
$u(P_r^m) < 0 < u(P_s^m)$
$u(P_r^m) + u(P_s^m) = 0$

* $P_s^m (\& P_r^h)$

Hidden characteristics:
none

4.19 Determining the market equilibrium (with profit maximisation)
If we look at the same market but change the market structure assumption to T2, what will happen then? Positive utility is provided by P_s^h and P_s^m. But, since T2 means maximising profit whilst keeping the $\mathcal{B}_s$-type at its reservation level, only P_s^m remains. In addition P_r^h can be offered, but it only increases sales volume and does not help to further increase utility.

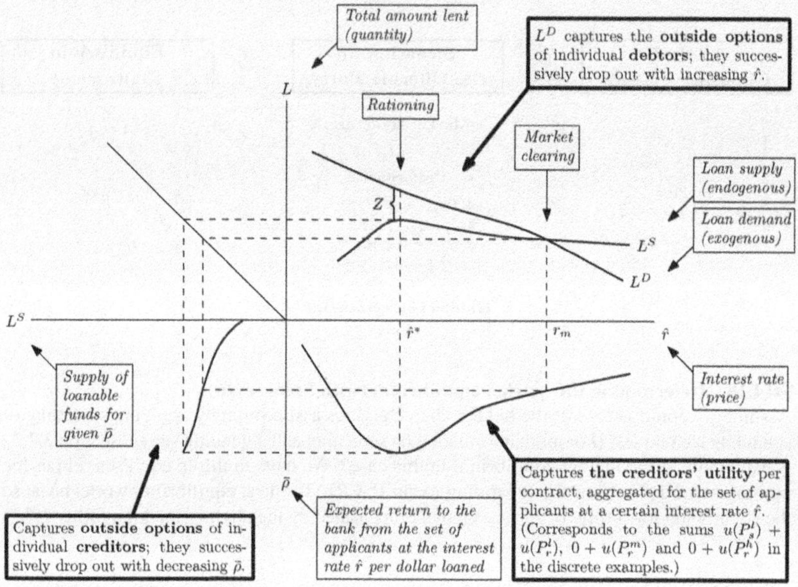

4.20 Determining the market equilibrium ([SW81] example)
The examples so far show the theory of market mechanics in the simplest form possible. But of course we can also construct continuous examples. One sample is provided by Stiglitz and Weiss in [SW81]. Remark: The graph is a reproduction of figure 4 "Determination of the Market Equilibrium" in [SW81, p. 397]. The boxes with arrows have been added to supplement the graph with my own explanations. Note that in reality the exact functions may look different than depicted by Stiglitz and Weiss, compare [AR09]. However, the crucial point here is to demonstrate in principle how the market equilibrium is determined.

To reinforce the modular mindset one more time: How does the example in figure 4.20 connect to the evaluation module? The lender's utility function per individual contract could come from a situation like the one in example 4.13. The utility function calculated there tells the expected value from a certain contract. To calculate the mean return $\bar{\rho}$ in figure 4.20, these function need to be aggregated for the pool of borrowers applying at a certain $\hat{r}$ – like we did in the discrete example δ when calculating $u(P_s^l) + u(P_r^l)$, $0 + u(P_r^m)$ and $0 + u(P_r^h)$. We need to know which borrowers apply at a certain $\hat{r}$ and what the utility of contracting with them is.[63] Thus the aggregate function will heavily depend on the construction of the pool of borrowers and assumptions on outside options.

63 Sums must be weighted if the types in the pool of applying players have different cardinality.

Another continuous example is the self-selection principle presented in [Bes85]. Recall figure 3.20 in subsection 3.7.3. The indifference curves represent the preferences of the borrowers and the self-selection equilibrium mimics that of figure 4.17 above. If we add the indifference curves for the lender (one utility function per borrower as they are assumed to have different abilities to repay) and make assumptions about the market structure (T1 or T2), we can determine the exact equilibrium contracts (see figure 2 in [Bes85, p. 853]).

What is the take-away from all the above examples on different market mechanics? There are two things to keep in mind! First, let's have a look at the volatility of the market equilibrium. Notice that in all the above market models the products did not change at all. The assumption about the market structure changed. The assumption about available outside options changed. The assumption about the information distribution changed. And when the preferences changed from figure 4.16 to 4.17 this could be due to changing assumptions in the evaluation module. The products, however, remained pretty untouched. Nevertheless, we encountered a variety of different equilibria. We observed how a market with the same products can lead to various different market equilibria. Hence, our first take-away: The volatility of market equilibria produced by different market constructions is immense![64]
Second, let's reflect on the degree of abstraction and generality provided by the contracting module. The products in the example above could have simply been numbered. Not knowing what they stand for is, of course, no problem for analysing the market equilibrium. But the symbols and indices have been chosen to be a bit suggestive. We can think of different interpretations.

- P_s^l could represent a credit contract with loan terms that combine a save ability to repay (s) and a low interest rate (l). In the examples above we assumed $u(P_s^l) = 0$ which could indicate that such contract leads to break-even on the part of the offering party. The fact that the $\mathcal{B}_s$-type only has access to the products P_s^l, P_s^m, P_s^h could be interpreted as a borrower that cannot choose his ability to repay freely. The whole market can be seen with this interpretation in mind.
- P_s^l could also represent a sales contract in the market for lemons, where a super good car (s) is sold with a low markdown (l). P_r^h could represent a sales contract where a rotten car (r) is sold with a high markdown (h). The informed players are the current car owners, the uninformed party is some car dealer offering a certain amount of money in return for a second-hand car.

64 Therefore it makes perfect sense not to spoil clear results in the contract module with arbitrary assumptions about a fictive market. We should keep contract mechanics and market mechanics separated and only combine it for concrete applications.

Market mechanics are not credit contract specific. They can be researched on their own. Even in the continuous examples: the construction of the rationing equilibrium in figure 4.20 can be understood without the information that it is all about credit contracts, and likewise figure 3.20 is abstract enough to replace the axes by the properties of any other contract or any other product. Therefore, our second take-away: Obviously, we can analyse markets at an abstract level without reference to a particular product. Moreover, the same market model can be applied to different markets (e.g. market for credit contracts and market for lemons in the examples above). This again demonstrates how the modular approach – in addition to providing structural clarity and complexity control – increases research efficiency by avoiding redundancy.

4.4 Assembling and disconnecting modules

4.4.1 The product layer module chain

Recall that on the product layer we work at a level of abstraction where we pretty much forget what product the theory is all about. Whether it is credit contracts, insurance contracts, second-hand cars or any other contract does not matter at this layer. Thereby we can talk about, discuss and research these things in a universal manner. Also the results acquired at this layer will be quite general and thereby be applicable to a wide variety of different situations.

In this chapter we have structured the product layer into three modules. Figure 4.21 presents a nice overview. As we have seen, each module can do a variety of different things; e.g. the evaluation module performs different transformations depending on the evaluation criterion applied. These modulations of a module's transformation are captured by a SELECTION-INPUT. The interdependencies between consecutive modules are captured by the respective interfaces DATA-INPUT and DATA-OUTPUT.[65]

What happens at the product layer of lending theory in general are the following steps. First, we start with a pile of different contracts. Then each player sorts this pile according to his preferences. Having established their preference relation they finally play the contracting game and settle on a certain contract.

65 The wording is loosely based on terms borrowed from computer science. E.g. compare the selection-bit of a demultiplexer chip.

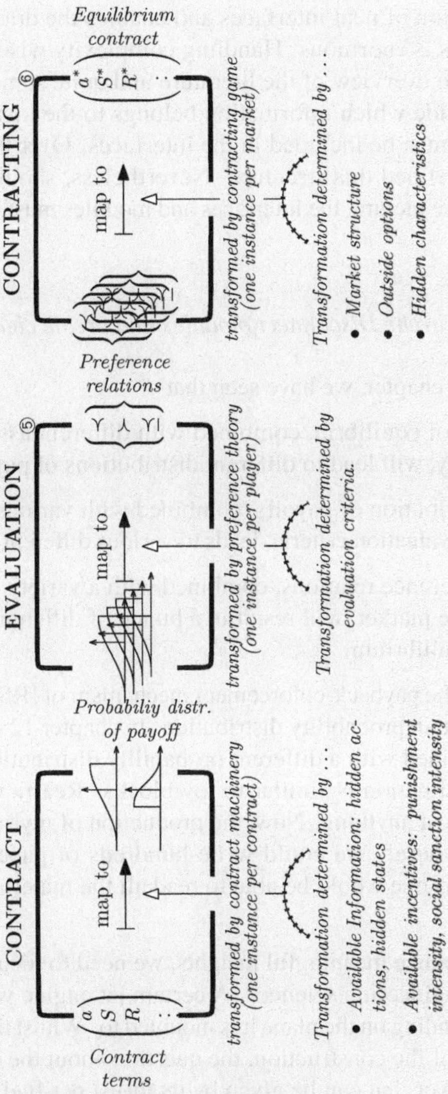

4.21 Combination of module boxes (with modulators)
A schematic illustration of the layers of abstraction and their interplay. The interfaces of a module consist of a DATA-INPUT, DATA-OUTPUT and SELECTION-INPUT. The SELECTION-INPUT contains additional assumptions to determine how the transformation from DATA-INPUT to DATA-OUTPUT in performed.

Don't be misled by the simplicity of figure 4.21. The intellectual challenge that precedes the definition of neat interfaces and enables the drawing of such beautifully clear structures is enormous. Handling complexity wisely is challenging. It requires an extensive overview of the literature and quite some thinking to extract the essence and decide which information belongs to the inside of a module and which information must be included in the interfaces. Over the last half decade I have continuously refined this structure. Nevertheless, should interdependences arise that do not fit the picture, the interfaces and modules must be further sharpened up.

4.4.2 *For basic research: Disconnect modules and create clear-cut results*

In the course of this chapter, we have seen that

- a single map of equilibria, combined with different assumptions about the ability to repay, will lead to different distributions of payoffs,
- the same distribution of payoffs, combined with various assumptions about the subject's evaluation criteria, leads to various different preference relations,
- the same preference relations, combined with a variety of different assumptions about the market, will result in a bunch of different results concerning the market equilibrium.

Another example is the payback-enforcement mechanism of [RS04] which is already combined with a certain probability distribution. In chapter 12 of this dissertation it is shown that, combined with a different probability distribution, its quality drops considerably. The situation is similar to toy blocks: Rearrange the basic blocks and you can get almost anything. Now, the production of myriads of combinations doesn't get us anywhere! We could write hundreds of papers with all sorts of combinations, but no-one would be able to read all the material and aggregate the results.

Consequently, to produce meaningful insights, we need to focus. Let's consider an example from the engineering sciences. A certain jet engine will achieve different cruising speeds depending on the plane it is mounted to. Whilst the speed of the plane depends on the rest of the construction, the question about the engine performance has a satisfying answer that can be given by its thrust per fuel flow. Similarly, the performance of the cross-reporting mechanism invented in section 12.8 – since developed independently from other modules – is a solid result. It *always* provides zero equilibrium deadweight loss. That's for sure. Another example: It will never be possible to exceed the repayment rate of the joint liability contract with an adaptive punishment regimen developed in chapter 10. That's a fact. It *always* extracts the maximum possible repayment. Overall, we can talk about maps of repayment, payback enforcement, efficiency etc. much more precisely without blurring it by arbitrary assumptions about specific market situations. Likewise, we can talk about market phenomena in a universal way without the clutter on the details of the actual product. Therefore, within the scope of basic research, it is rational to disconnect the modules and then create clear-cut results.

This dissertation is aimed at delivering basic research. As the title suggests, the focus lies on improving the design of credit contracts. Thus it does not talk much about market mechanics. What might feel unfinished to monolithic modellers is a purposeful decision. In fact, all the structural work in part I. of this dissertation is aimed at preparing the modular course of action in part II.

Subsequently, this dissertation puts a particular emphasis on the improvement of payback-enforcement schemes. The algorithm in appendix A has especially been developed to analyse these. In a way, we are improving the jet engine that boosts credit contracts – the payback-enforcement module. The catalogue of repayment games (figure 13.2) depicts the applicability of each scheme and indicates its performance (equilibrium repayment per equilibrium punishment load).

4.4.3 For applications: Assemble modules and produce custom-made recommendations

What are the advantages of a modular approach for applications? Consider the paper of Arnold, Reeder, Steger [ARS13]. They take the repayment game from [BC95] [payback-enforcement module], combine it with uniformly distributed marginals and the independence copula [ability to repay], evaluate the resulting distributions of payoff from the creditor's perspective (criterion: expected repayment) and the debtor's perspetive (criterion: expected utility) [evaluation module], and then put the whole thing in a T1 market structure with hidden characteristics [contracting module]. Unfortunately, we do not know how the results change if we apply a more realistic triangular distribution[66] or what happens if the market structure is changed to T2. What if the borrowers are risk-friendly or the the lender is risk-averse (compare infobox 3.10 on p. 68)? As a matter of fact, we cannot use the results to make recommendations for even slightly different situations. Without a match to a given concrete real-life situation the assumptions appear rather arbitrary. In general, the results of monolithic models are extremely specific and hardly applicable.

66 Extreme realisations are usually less likely.

With the modular approach, however, we can quickly adopt our models to diverse real-life situations and provide tailor-made recommendations by recombining existing knowledge from the individual modules. Moreover, we can give advice on different levels. The practical advices given in this dissertation are mostly directed to a loan officer being faced with a certain payback-enforcement problem. The audience of researchers focusing on market mechanics, however, are politicians and CEOs of financial institutions.

So, in addition to facilitating basic research, the modular approach also comes along with better applicability. In fact, it lifts the quality of our policy suggestions to a whole new level where we can easily provide tailor-made recommendations for a whole variety of different situations. "Specific models are generalized because general models can be more effective. But in individual applications it is advantageous to make general models specific again." [DS63, p. 34] That's exactly what the modular approach facilitates: The generation of general results (within individual modules) and the specification of the theory to concrete applications (by recombining the modules adequately).

I'm not an expert on stress tests of banks, but I am conviced that the simulations should include changes in the evaluation criteria of subjects as well as chanching outside options and market structures. The modular approach may, for example, be useful to help simulating all these influences in a structured fashion.

This dissertation is focused on credit contract design. Nevertheless, the interactions with neighbouring modules are discussed and practical advices for the application of the results are given, too:

- sections 6.3 and 6.4 in chapter 6,
- section 7.5 in chapter 7,
- practical advice at the end of chapter 10,
- practical advice at the end of chapter 11,
- practical advices in 12.3.1, in 12.3.2 and at the end of 12.4.

5 Standardisation: The standardisation of credit contract theory

So far we have discussed the possibility to and the benefits of organising lending theory in modules. This section will analyse some general structures of lending theory, unrelated to particular modules. The analysis will lead to a number of recommendations for how to improve lending theory by using uniform concepts and standards.

5.1 Flows of value in a credit contract

To structure the diversity of different concepts like interest, collateral, liability, penalties, cost of monitoring, pecuniary rewards, etc. a thorough analysis of flows of value in the lender-borrower-system is necessary. The detailed analysis of flows of value in a lending game is the cornerstone for the formalisation of payoff rules in 5.3, which in turn paves the way for the digitalisation of repayment games (see section A.1.4).

We can distinguish three fundamental types of flows of value.

Transfers, i.e. value that is transferred between subjects (e.g. repayment or transfer payments between peers in a group lending contract).

Losses, i.e. value that flows out of the system (e.g. monitoring costs, penalties).

Gains, i.e. value that enters the system (e.g. yields of an investment project).

In short: Value is either lost or gained or transferred.

We intentionally talk quite generally of flows of *value*, as these values can take different forms. It can be money, assets, or any other valuable resources. For example, interest is typically a transfer of money from borrower to lender whilst collateral often is a transfer of goods (e.g. a building) from borrower to lender. For many models, the concrete form of the value is rather irrelevant. It is the *direction* of a flow and the *amount* of value that matters when analysing incentives and determining optimal strategies! To catch the essence we utilise abstraction.

Figure 2.1 illustrates potential flows of value in a reciprocity agreement with thin arrows. Thereby sender, receiver and thus directions are specified. Whilst these

C. Prem, *The Theory of Credit Contracts*,
https://doi.org/10.1007/978-3-658-29362-8_5

flow-of-value-vectors are to some degree universal for every credit contract, the amount of money that flows along each edge remains unsettled for the moment (may even be zero) and is contract specific.

Let's walk briefly through these arrows in figure 2.1. Initially, cash (or any other valuable resource) flows from creditor to debtor. The debtor uses it one way or the other. So in models where purpose enforcement is an issue, this edge may be frayed out and split into several alternative paths to the 'nature player' $\mathcal{N}$. Next, $\mathcal{N}$ determines available resources at repayment time and sends it back to the borrower. Finally, the borrower decides how much he repays to the creditor and how much he keeps to his own, i.e. he splits available resources into his part and the lender's part.

In addition the need for enforcement may produce losses, i.e. value that leaves the lender-borrower-system (e.g. monitoring costs, auditing costs, punishments, or any other credit specific transaction costs). These resources are lost, burned up, sort of consumed by the credit contract. They are not transfers between lender and borrower, and are therefore not part of the actual reciprocity.

The 'outside of the system' can be imagined as a kind of third artificial entity. Then gains and losses are transfers to or from this external entity. Thereby we get some 'law of conservation of value', like in physics. Put in game theory language, the game becomes a zero-sum game.[67] And by doing so the welfare effects of a contract can be taken as a reading from the utility of this external entity – with opposite sign, of course.

To summarise the insights of this subsection:

- We have abstracted from concrete forms of value to just value, since the form does not matter for the analysis of lending games.
- Subsequently, we focused on the direction and size of the flows of value.
- Concerning directions we distinguished transfers from gains and losses.

5.2 Timelines and design principles

In this subsection we take a look at the chronological construction of lending models. Timing is important! For example, concerning the problem of diversion of funds, it makes a difference if the monitoring intensity decision is made before or after the investment decision. And since timing is important, in their papers authors often provide a timeline by some means or other.

67 Such outside-entity is of course a passive 'player' with empty action sets.

5.1 Specification of timelines in the literature

[Arm99] **"On the design of a credit agreement with peer monitoring"** (p. 84)
1 "contracting and investment"
2 "monitoring decisions"
3 "return realizations & borrowers' repayment/default decisions"
4 "monitors learn about about each others' return realizations"
5 "social sanctions and non-refinancing penalties"

[KV00] **"Optimal Contracts when Enforcement is a Decision Variable"** (p. 122)
1 "nature moves first and chooses $x \in X$ at time 1"
2 "voluntary payment v to make at time 2"
3 "whether to go to court to enforce F at time 3"

[AG00] **"Peer group formation in an adverse selection model"** (p. 636)
1 "At time 1, borrowers form random pairs."
2 "At time 2, each pair decides whether or not to request a loan."
3 "At time 3, banks compete à la Bertrand to finance the pairs ..."
4 "At time 4, production takes place. Nature decides about return realisations, and the banks are reimbursed according to the joint liability clause."

[RS04] **"Is Grameen Lending Efficient? Repayment Incentives and Insurance in Village Economies"** (p. 221)
1 "Invest"
2 "Output y_i"
3 "Repay b_i, Message m_i, Refund z_i, Punishment q_i"

[BSW14] **"Group Lending and Endogenous Social Sanctions"** (p. 5–6)
1 "The bank offers a menu of loan contracts, which specify loan size, interest rate, whether the loans entail individual or joint liability and the penalty for the individual or the group for defaulting on the loan;"
2 "The potential borrowers choose which loan contract to take if any;"
3 "If a loan is taken, the borrower decides how to invest it;"
4 "Once project returns are realised, borrowers decide whether or not to repay the bank;"
5 "Finally, from observing their repayment behaviour, the community decides whether or not to include borrowers in a communal activity."

[DQFG16] **"Group lending without joint liability"** (p. 220)
1 "Borrowers form pairs, and agree on a repayment rule."
2 "Loans are disbursed, borrowers observe the state and simultaneously make repayments (the repayment game)."
3 "Conditional on repayments, contracts are renewed or terminated and social sanctions are carried out."
4 "If an IL borrower's partner was terminated but she repaid, she rematches with a new partner."

To get an impression of the landscape of different timelines I gathered a collection of quotes of how authors describe the timing in their lending models (see p. 105). At first glance, it feels like a complex and unclear variety of different timelines with little similarities. The reason is that timelines in the literature are far from standardised. Thus the similarities and differences are not coming to light easily.

Therefore, let's try to set up reasonable quality standards for timelines. Basically, a timeline is a sequence of events! So the first question is: Which events can occur? With the examples above we already have a reasonable set of examples to work with. But can we provide some typification of these events to structure the bulk?

The most important type of events are certainly *decision nodes* of the players, i.e. first and foremost creditors and debtors. Random events can be considered as decision nodes of a nature player. Furthermore we have *deterministic events* like the disbursement of a loan after contracting, the execution of predefined penalties and sanctions, or events where some information becomes known to a player. Of course, we can distinguish a variety of subtypes whenever relevant e.g. decision nodes with continuous or discrete action sets, or deterministic events that reveal information (*information events*) versus events that trigger flows of value (*payoff events*), etc.

The type of an event is determined by the modeller. For example, in many models penalties and social sanctions are deterministic events. But there are other models in which they are decision nodes, e.g. [KV00, BSW14], see p. 105. In general, making variables endogenous means replacing a deterministic event with a decision node. The necessary construction of the action space at the new decision node is the reason that this process can result in different models.

A brief look back, at the examples given on p. 105, reveals a rather sloppy handling of events in the literature. We encounter the following conceptual problems.

- Occasionally, there are cases where events are non-atomic, e.g. contracting in (1) of [Arm99] can be a quite complex sub-game, likewise forming pairs in (1) of [DQFG16]. What is the procedure for that? A proper specification of contracting can be found in (1-2) of [BSW14], likewise [AG00] (1) specifies how pairs are formed.
- Sometimes the player of a decision node remains unclear, e.g. in (2) of [AG00]. Does the pair act as a new homogeneous player?
- There are instances where several events are assigned to the same point in time. Whilst the first idea is to interpret this as an indication for simultaneity, in some cases this interpretation does make absolutely no sense (e.g. (1) of [Arm99], (3) of [RS04]).

So, how can we improve the quality of timelines? A precise timeline sticks to the following design principles.

1. **Chronology** t. The first column indicates the point in time.[68] If we want to indicate the simultaneity of two events, both are assigned to the same t, and for a graphical indication the two rows may be separated by a dashed line instead of a solid one.[69]

2. **Players** $\mathcal{P}$. The second column determines the owner of a decision node ($\mathcal{C}$ creditor, $\mathcal{D}$ debtor, $\mathcal{N}$ nature). No entry marks a deterministic events.

3. **Description/Action space**. The third column provides a description of the event and, in case of decision nodes, the action space of the player. Standard events of credit contracts should be referred to by their standard names.[70]

Let's apply these standards to the timeline in [Arm99] (see figure 5.1).

How do we handle the specification of varying information distributions in the course of the timeline? Which player has which information at which decision point? Does he also have the information about what information the other players have? The problem is the sheer enormity of possible information. To avoid having to treat each information individually we should establish some default convention. Whilst in reality it would be easier to specify the knowledge of a person instead of his ignorance, for lending theory it is the other way round. It is easier to say: everything is public knowledge, expect from the particularly mentioned information. Thus, regarding information, we supplement the rules for high-quality timelinesby the following additional one.

4. **Information distribution** at each point in time. By default we assume all information is publicly known to every player in the game. If we have an event that reveals some information to some player we implicitly know that this has been unknown to this particular player beforehand. Permanently hidden information that is never revealed by an information event does not appear in the timeline and must be specified separately.

In the model of [Arm99] presented on p. 108 it is therefore important to separately denote that the return realizations are unknown to the creditor throughout the game, whilst it is implicitly clear that the information is unknown to the other borrowers until $t = 9$.

68 Note that these are discrete points. Consequently, there is no meaning in the phrase 'at the end of the period'! Two events either happen simultaneously or successively.

69 The suggestion to separate rows with simultaneous events by dashed lines mimics the convention in extensive-form games to connect decision nodes from the same information set with dashed lines.

70 In figure 5.1 the standard names are supplemented in square brackets.

t	$\mathcal{P}$	Description/Action space
1	$\mathcal{C}$	**Contract offer** The creditor sets the contract terms: $\bar{R} \in \mathbb{R}^+$ gross interest $\eta \in [0,1]$ joint liability fraction $\beta \in [0,1]$ punishment intensity (modelled as punishm. probability)
2	–	**Contract acceptance** Borrowers accept the loan (no bargaining modelled).
3	–	**Disbursement [Service]**
4	–	**Investment [Provisions for repayment]** No diversion of funds or effort levels modelled.
5	$\mathcal{D}$	**Auditing decision [Information part of enforcement]** $\gamma \in [0,1]$ Auditing intensity (modelled as auditing probability).
6	$\mathcal{N}$	**Return realizations [Ability to repay]** $(\theta_1, \theta_2) \in \{0, \bar{\theta}\}^2$ Each borrower is either successful or fortuneless. Bernoulli distribution, probability of success α, uncorrelated between borrowers.
7	$\mathcal{D}$	**Repayment decision [Willingness to repay]** $\rho \in [0,1]$ Probability of strategic default (mixed strategy decision point).
8	$\mathcal{N}$	**Repayment [Return service]** $\vartheta \in \{0,1\}$ Strategic default or repayment. RV with Bernoulli distribution, probability of repayment ρ (see $t=7$).
9	$\mathcal{N}$	**Auditing result** $\tau \in \{0,1\}$ Borrowers learn about each others' return realizations. RV with Bernoulli distr., revelation prob. γ (see $t=5$).
10	–	**Punishment & Social sanctions [Incentive part of enforcement]** Defaulting groups are punished with probability β (see $t=1$), $-V$ each. Unmasked strategically defaulting borrowers suffer social sanctions W.

5.1 The timeline in [Arm99]
This example demonstrates how the complex model of Armendáriz can concisely be captured by sticking to the design principles for timelines introduced on p. 107. Remarks:
$t=5$: They use the term monitoring, but they actually mean auditing (see p. 27). Also, the paper only considers symmetric equilibria (see [Arm99, p. 84]), thus no notational distinction between $\mathcal{D}_1$ and $\mathcal{D}_2$ and potentially different actions γ_1, γ_2.
$t=7$: Likewise, "ρ denote[s] the probability of strategic default by either borrower in a symmetric equilibrium." [Arm99, p. 84], there is no distinction between the borrowers.
$t=8$: To give the random variable a name, ϑ has been added. It remains unclear if in their model the random repayment or strategic default decision is the same for both borrowers or whether this random variable is meant to be a 2-dimensional one. No information about any dependence structure is given in the paper. When they compare payoff from strategic default with the payoff without strategic default in inequality (3) on page 85, the case of a strategically defaulting partner is not even considered. If this random variable is 1-dimensional it should be considered on the right hand side. If this random variable is 2-dimensional, it should be considered on both sides weighted with probability $\alpha \cdot \rho$ for a successful partner that strategically defaults, respectively.
$t=9$: To give the random variable a name, τ has been added.

Q 5.1 Timelines vs. extensive-form and normal-form games

The purpose of the tool 'timeline' is the description of a lending game. But, we already have tools to specify games, namely extensive-form and normal-form! They comprise everything: the chronology, decision nodes, its players and their action sets, the information distribution and eventually the payoff of each player for every outcome of the game. Unfortunately they become unwieldy for comprehensive lending games like in [Arm99]. So – apart from special cases like [BC95, Fig.1&2] which focus on a very small detail (the repayment game) only – drawing the extensive-form game tree is not practicable.
In contrast, timelines provide an alternative possibility to describe a game. They are, so to speak, compressed versions of game trees! Nevertheless, they must be as precise as extensive-form or normal-form games. By knowing the timeline it must be possible to unambiguously reproduce the game.

Q 5.2 Timelines with switch points

Timelines can be problematic, if – after a decision node – the game continues in different ways, depending on the action of the respective player. Fortunately, these are often dead-ends and can be complemented by common sense. For instance, if one party rejects the contract during the contracting phase, the game is at its end. And if the ability to repay is zero, the repayment decision is meaningless. In more complex cases the timeline can be split into several tracks at such decision points, like a fork.

Finally, let's ponder how events can be permuted. Which chronological restrictions do we face? Naturally, some events will always have to happen in a certain order. Most importantly the standard events service, provisions for repayment, realisation of the ability to repay, repayment decision and repayment will always appear in exactly this chronological order. But other parts of the contract can be placed in the timeline with some freedom. For example the auditing intensity decision could theoretically be placed before investment, after investment or after return realizations or even after repayment. The auditing result could be placed before repayment or after repayment. However, the auditing result can only occur after the auditing intensity decision. So from a chronological perspective some irrevocable fixed sequences are surrounded by a bit of flexibility.

What is the actual difference between models with different timing, what is the effect of swapping events? It is the information that is available at a certain decision point that is changed if the timing of events is changed. In short: The chronology of events determines the information distribution at each decision point.

Although the timing of events is crucial, timeline variations are sometimes only discussed on the sidelines of papers. For example in footnote 13 of [Arm99, p. 84] the author simply states that a different order "would complicate the analysis, without adding new insights."

5.3 Payoff rules and redundancy in payoff functions

A timeline is a condensed version of an extensive form game tree, but what's missing are the payoffs at the leaves of this tree. Specifying the payoffs for each outcome can become laborious. For each player it is a function whose domain is the cartesian product of all action spaces occurring in the game.

Fortunately, there is quite some redundancy in the payoffs. The redundancy is unveiled if we focus on the general components of a player's payoff rather than at the single leaves of the game tree! These components originate from the flows of value (see section 5.1) that are triggered by the decisions at certain decision nodes.[71]

So how can we specify the payoffs for each outcome of the game in the most concise manner? We can do that by writing down the conditions that trigger a certain flow of value. Figure 5.2 demonstrates how the flows can be formally stated in case of the model of [Arm99]. Then a debtor's payoff in each situation of the game can be easily derived by adding the relevant intensities in the last column. Similarly, the creditor's payoff consists of all transfers $\mathcal{D} \rightarrow \mathcal{C}$ minus all transfers $\mathcal{C} \rightarrow \mathcal{D}$ (the bank carries no gains or losses in this model).

Eventually, to complete the specification of a lending game we need one more rule.

5. **Payoffs** for each outcome of the game. The best practice is to split the payoff into components (e.g. auditing costs, transfer payments, social sanctions) and specify the conditions under which each component applies.

Q 5.3 A new standard for specifying lending models?

Note, how the whole model of [Arm99] is basically captured by the timeline (figure 5.1) and the flows of value (figure 5.2) in just two pages. It contains all the necessary ingredients to specify a game: players, actions, payoffs, information (known as the acronym "PAPI", see [Ras07]). It is clearly arranged, easy to grasp and more precise than several pages of running text. Moreover it is formal enough to be understood by a computer, without further ado the rules can be coded into a software algorithm. If every paper were presented in this fashion, reading, understanding and above all the comparison of different models would be nice and easy.

71 Not all decision nodes are involved in the steering of flows of value. Some determine flows of information, e.g. $t = 7$ in figure 5.1.

Component [Type]	Rule [Decision node]	Intensity
Disbursed amount [transfer $\mathcal{C} \to \mathcal{D}$]	always applies [$t = 3$]	$+K$
Investment [loss for $\mathcal{D}$]	always applies [$t = 4$]	$-K$
Cost of monitoring [loss for $\mathcal{D}$]	always applies [$t = 5$]	$-c\gamma$
Project return [gain for $\mathcal{D}$]	applies if $\theta_1 = \bar{\theta}$ [$t = 6$]	$\bar{\theta}$
Repayment [transfer $\mathcal{D} \to \mathcal{C}$]	applies if $\theta_1 = \bar{\theta}$ [$t = 6$] AND no strategic default [$t = 8$]	$-\bar{R}$
Liability [transfer $\mathcal{D} \to \mathcal{C}$]	applies if $(\theta_1, \theta_2) = (\bar{\theta}, 0)$ [$t = 6$] AND no strategic default [$t = 8$]	$-\eta\bar{R}$
Future financing [gain for $\mathcal{D}$]	applies if $(\theta_1, \theta_2) \neq (0,0)$ [$t = 6$] AND no strategic default [$t = 8$]	V
Reduced financing [gain for $\mathcal{D}$]	applies if $(\theta_1, \theta_2) = (0,0)$ [$t = 6$] OR strategic default [$t = 8$]	$(1-\beta)V$
Social sanctions [loss for $\mathcal{D}$]	applies if $(\theta_1, \theta_2) = (\bar{\theta}, 0)$ [$t = 6$] AND strategic default [$t = 8$] AND revealed return realizations [$t = 9$]	$-W$

5.2 Flows of value in [Arm99].
This tabular demonstrates how the flows in a lending game are expressed with formal rules. The types in square brackets tell the type of a flow of value, see p. 103 ff. It becomes obvious how the flows depend on the various decision nodes and decisions (compare p. 108).

5.4 Symbols

Reading and comparing different models is easier if they all speak the same language. Thus, to conclude the analysis of general structures of lending theory, a few words on the notation of lending models.

Q 5.4 The tools of notation

When choosing a notation we have the following tools at hand:

- A symbol consists of a main symbol (e.g. a latin, greek or cyrillic letter).
- If required additional marks ($^\circ$, $*$, ', $\tilde{}$, $\hat{}$, $\check{}$, $\vec{}$, ...) or indices can be added.
- Most symbols come in two versions as a capital letter or minuscule (e.g. A, a, Δ, δ, ...).
- The typeface style can be varied (italic C, bold $\mathbf{C}$, calligraphic $\mathcal{C}$, blackboard bold $\mathbb{C}$, ...).

Symbol and typography often give a hint at the meaning of the symbol. Additional marks and indices are additive and can provide helpful information on the type of a variable. E.g. $\vec{v}$ usually indicates that v is a vector.

Inventing a good notation is not easy, as good notation must reflect the structure of the theory! Therefore general structures must be analysed and thoroughly understood first. Good notation reveals as much as possible about its meaning and is at best self-explanatory, thus creating a strong link between semantic and syntactic. Ideally it would be possible to read and understand a model without additional explanations on notation. In physics this works quite well. Most often we write m for mass, v for velocity, V for volume, F for force, and so on.[72]

Q 5.5 The ambition for good notation in other fields

Chemists have gone through several decades of systematisation attempts to structure chemical elements before Dmitri Mendeleev invented the well-known modern periodic table. The benefit? Firstly, by sorting elements in rows and columns according to their atomic number it became obvious that the characteristics of elements repeat in regular patterns. Elements could be structured into the major categories metals, metalloids and nonmetals, according to their physical and chemical properties. Second, gaps in the table revealed gaps in mankind's knowledge about the earth's elements and led to the discovery of – until then – unknown elements like e.g. gallium and germanium.
Structural work is laborious and requires fastidious researchers! However, it is a fundamental part of the cognition process. Likewise, the useful and beautiful Leibniz notation in calculus is the result of several years of adjusting and discussion.

72 There even exist standards on notation, e.g. DIN 1304 on formula symbols or the "International System of Quantities" standard ISO 80000.

In lending theory, however, we encounter a bewildering notational plurality. It is a symptom of the rather unsystematic exemplifying research in the subject. It indicates that – until now – there has not been put much effort in carving out the general structures of lending theory. It also shows that a joint research strategy is missing. Page 114 presents some examples from the literature.

This inconsistency concerning the symbols used is an annoyance as it hinders jumping between papers fluently. Instead, you have to work yourself into the individual notation of a paper each time you want to look something up. Not even the return service – a fundamental part of every credit contract – has a standard symbol. I started writing notation tabulars to be able to re-read papers without having to search the text for the meaning of particular symbols again and again. Notational inconsistency impedes efficient research.

To improve this proliferation it would be nice to have some task force that elaborates and publishes a code of practice for the presentation of lending models, including guidelines for the use of symbols, as well as standards for the notation of timelines and payoff rules.

Subsequently, the production of a catalogue of all existing lending models using the proposed standard notation could serve the following three purposes:

1. A catalogue of all lending models could serve as touchstone for the notational standards defined, to test if they are applicable and practicable.
2. It would fix the state of the art and present today's knowledge in a well-structured compact fashion.
3. It may very likely trigger further research as such catalogue has the potential to reveal blank areas in our knowledge about lending.

It is self-evident that notation influences a researchers everyday work. A well thought through notation has enormous potential. On the other hand, notation can obstruct a researchers view on his subject. Chapter 5 should be seen as the first attempt towards a standardisation of lending theory. With good notation we will understand lending theory better and research faster.

5.2 Symbols and notation used in the literature

[Dia84]	**"Financial Intermediation and Delegated Monitoring"**
	[Ability to Repay]
$\tilde{y}$	random variable for "the total output of the project" (p. 396)
	[Return Service]
z	"aggregate payment which the entrepreneur pays to the m lender" (p. 396)
	[Payback-Incentives ↯]
ϕ	"non-pecuniary penalty function" (p. 397)
[BC95]	**"Group lending, repayment incentives and social collateral"**
	[Ability to Repay]
$F(\theta)$	CDF of probability distribution (see p. 3)
$[\underline{\theta}, \overline{\theta}]$	domain of CDF $F(\theta)$ (see p. 3)
	[Return Service]
r	"amount to be repaid, inclusive of interest" (p. 4)
$2r$	amount due in case of bail out (see p. 5)
	[Payback-Incentives ↯]
$p(\theta)$	the penalty the bank can impose on the defaulting borrower (see p. 4)
$s(L, \theta)$	"the social penalty function" (p. 10)
[Arm99]	**"On the design of a credit agreement with peer monitoring"**
	[Ability to Repay]
α	probability of success (see p. 82)
$\{\underline{\theta}, \overline{\theta}\}$	possible return realizations (see p. 82)
	[Return Service]
$\overline{R}$	repayment to the bank when peer is fortunate (see p. 85)
$(1+\eta)\overline{R}$	repayment when peer is unfortunate (see p. 83, 85)
	[Payback-Incentives ↯]
V	bank sanctions, value of future financing (see p. 82)
W	social sanctions by peer (see p. 83)
[RS04]	**"Is Grameen Lending Efficient? Repayment Incentives and ..."**
	[Ability to Repay]
$p(y)$	PDF of investment output, i.e. "probability that state y occurs" (p. 220)
$\{0, h\}$	possible output realisations (see p. 220)
	[Return Service]
$b_i(y)$	"the payment from agent i to the bank in state y" (p. 220)
	[Payback-Incentives ↯]
$q_i(y)$	punishment the bank imposes on agent i in state y (see p. 220)
[All16]	**"Optimal (partial) group liability in microfinance lending"**
	[Ability to Repay]
R_i	"realized value of their respective returns to borrowing" (p. 203)
$\pi_{R,R}(P)$	"probability that the realized returns are such that both ... repay" (p. 204)
$\pi_{R,D}(P)$	probability that realized returns are such that partner defaults (see p. 204)
	[Return Service]
I	"cost of repaying the loan", including principal and interest (p. 203)
T	net transfer made from borrower to his partner (see p. 203)
	[Payback-Incentives ↯]
V	"present discounted value of being eligible to borrow" (p. 203)
P	monetary penalty the bank imposes for not bailing partner out (see p. 203)

Q 5.6 First proposal for a standardised notation in lending theory

- Standard symbols for players: $\mathcal{C}$ creditor, $\mathcal{D}$ debtor, $\mathcal{N}$ nature [calligraphic]
- Endogenous variables: x° Endogenous variables in equilibrium: x^{*}
- Transfers $\mathcal{C} \rightarrow \mathcal{D}$: $\acute{x}$ Transfers $\mathcal{C} \leftarrow \mathcal{D}$: $\grave{x}$ Transfers $\mathcal{D} \leftrightarrow \mathcal{D}$: $\tilde{x}$ (see p. 103)
- Standard transfers: $\acute{S}$ service (size of the loan), $\grave{R}$ return service (gross repayment), $\grave{C}$ collateral, $\grave{L}$ joint liability
- Gains: $\hat{x}$ Losses: $\check{x}$ (see p. 103)
- Standard losses: $\check{C}_S$ costs of screening, $\check{C}_M$ costs of monitoring, $\check{C}_A$ costs of auditing (costly state verification), $\check{P}_C$ punishments by the creditor, $\check{P}_{\mathcal{D}}$ (social) sanctions by borrowers
- Constants: X [upper-case character] Functions: $x(.)$ [lower-case character]
- etc.

Part II

Contributions to Selected Topics

6 Joint liability can cause adverse selection instead of resolving it

Abstract How does joint liability affect financial inclusion? Armendáriz and Morduch repeatedly argue that it makes credit cheaper (in terms of nominal interest rates) and can therefore "lure deserving safe types [of borrowers] back into the market." [AM10, p. 104] In contrast, I show that joint liability is expensive (in terms of expected repayment, maximum repayment and risk) and that in fact *less* borrowers will be eligible for credit.

6.1 How group lending went from euphoria to disillusion

The ideas of microfinance have not been new at all. In the 1850s already, credit cooperatives were firstly introduced in Germany, and can be seen as "forerunners of lending schemes that rely on joint liability." [GG99, p. 212] But the ideas grew increasingly popular with the Grameen Bank in Bangladesh and the International Year of Microcredit in 2005. It received an additional boost in popularity by Muhammad Yunus' Nobel Prize in 2006.

The fact that Yunus was not honoured with the Nobel Memorial Prize in Economic Sciences but with a Nobel Peace Prize provides evidence for the pronouncedly positive perception of group lending at that time. It was seen as the solution for credit market imperfections, as a tool for financial development, as a remedy against poverty.

After a phase of euphoria, however, the microfinance industry was confronted with increasing criticism for a number of reasons. The effect on poverty was disillusioning (see [HL07]), as reaching the poor with financial services is not the same as reducing poverty. It may even be the case that microfinance only makes the matter worse by driving borrowers into over-indebtedness (see [AB13]). The industry was also accused of a mission drift from serving the poor towards commercialisation (see [AM10, p. 239], [MS13]). Furthermore, group lending has negative social impacts as it – by design – leads to social pressure and public shame. Karlan finds "direct evidence that relationships deteriorate after default" [Kar07, p.F52]. It goes to such lengths that there are articles examining microfinance in conjunction with suicide (see [ASP15]).

C. Prem, *The Theory of Credit Contracts*,
https://doi.org/10.1007/978-3-658-29362-8_6

These drawbacks may be brushed aside as undesired side effects. An argument in favour of joint liability is that it reduces interest rates and therefore makes credit accessible to a larger quantity of borrowers. Armendáriz and Morduch, two experts in the field, repeatedly formulate it in their book "The Economics of Microfinance" [AM10]. This article will explain why that argumentation is flawed. Thereby, we are adding just another jigsaw piece to the doubts about joint liability – not at the list of side effects but at its very heart: the reciprocity agreement.

6.2 IL and JL contracts [reciprocity agreement]

The first part of the argumentation is situated entirely on the reciprocity layer (section 2.3) of lending theory. We replicate the assumptions in [AM10] and first of all describe what joint liability contracts (JL) and individual liability contracts (IL) do in terms of transfers.

As players there are a creditor $\mathcal{C}$, two identical borrowers $\mathcal{L}_1$ and $\mathcal{L}_2$, and a random player $\mathcal{N}$. Apart from the nature player $\mathcal{N}$, the decision points are all deterministic (p. 106) and moral hazard is not an issue. Since the borrowers always repay as stipulated, no enforcement modules are required. The timeline is presented in figure 6.1.

t	$\mathcal{P}$	Description/Action space
1	–	**Service** The creditor lends each of the two borrowers an amount of S.
2	$\mathcal{N}$	**Ability to repay** Possible realisations: $(a_1, a_2) \in \{0, A\}^2$ Marginals: Bernoulli distribution, probability of success π. Dependence structure: Independence copula. Assumption: The amount A is exogenously fixed and sufficiently high.
3	–	**Return service** Under IL borrowers repay according to the repayment function 0 for realisations $(a_1, a_2) = (0, \cdot)$ R_I for realisations $(a_1, a_2) = (A, \cdot)$ Under JL borrowers repay according to the repayment function 0 for realisations $(a_1, a_2) = (0, \cdot)$ R_J for realisations $(a_1, a_2) = (A, A)$ $2R_J$ for realisations $(a_1, a_2) = (A, 0)$

6.1 The timeline of the model
The simple model presented by Armendáriz and Morduch can be condensed into the following short timeline.

In this simple case, with only a single non-deterministic decision point, the payoff rules are straightforward and do not require an extra payoff tabular (section 5.3). The creditor's payoff function (i.e. his payoff in each outcome of the game) can be taken directly from the repayment functions (specified at $t = 3$ of the timeline) by adding $-S$ to the amount repaid ($-2S$ for JL, respectively). The debtors' payoff for each outcome of the game is the excess of his ability to repay A over his repayment. For each outcome it can also easily be read from the repayment functions specified at $t = 3$.

In the evaluation module we will work with the probability distributions of payoff (section 3.6) that a credit contract induces. Therefore we have to calculate these. They provide the interface that connects the contract layer to the product layer.

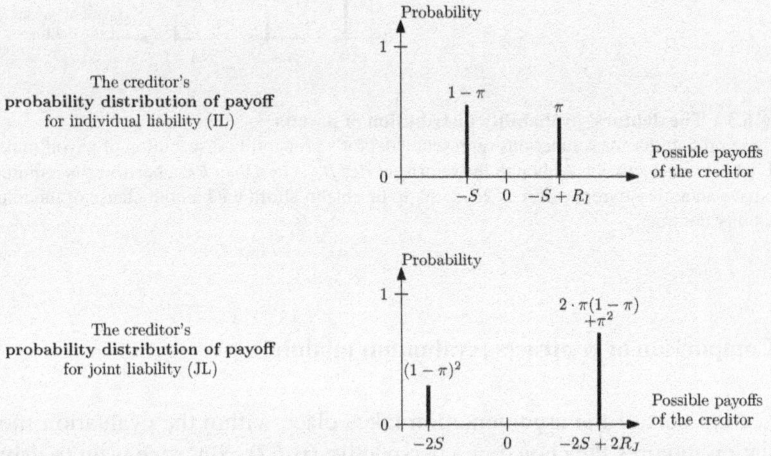

6.2 The creditor's probability distribution of payoffs
These probability mass functions represent the creditor's probability distribution of payoff under IL and JL, respectively.

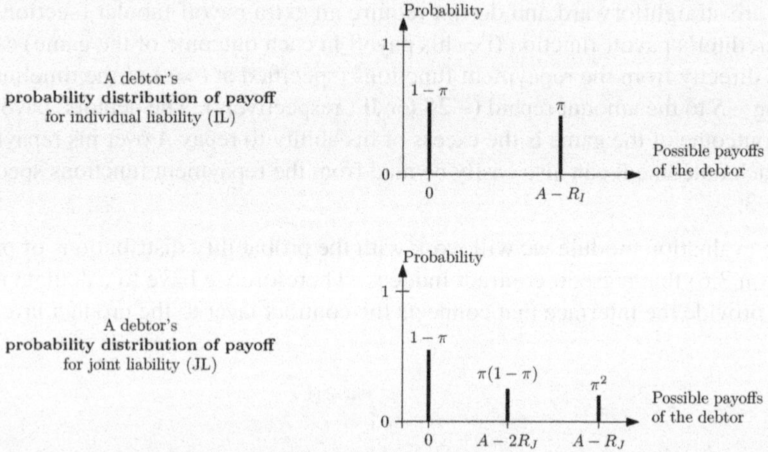

6.3 The debtors' probability distribution of payoffs
These probability mass functions represent a debtor's probability distribution of payoff under IL and JL, respectively. Although the return service R_J is less than R_I, a borrower is required to have an ability to repay of $A > 2R_J > R_I$ to be able to afford the bail out clause of the joint liability contract.

6.3 Comparison of contracts [evaluation module]

The second part of the argumentation takes place within the evaluation module. Evaluation requires the choice of a perspective (p. 67). Since we want to drive the topic towards issues like financial inclusion and adverse selection (rather than welfare considerations) the adequate choice is the creditor's and the debtor's perspective.

6.3.1 The creditor's perspective

First of all we need to decide about the criteria for comparison. In keeping with the model in [AM10] we make the assumption that the lender is placed in a T1 market structure (p. 68), which means that only contracts where the lender exactly breaks even come into consideration.[73] For each of the two contracts, IL and JL, the break-even condition can restated as a condition for gross interest R_I and R_J.

73 Less expected repayment lead to losses, more expected repayment goes under in competition.

	Break-even condition	Gross interest
IL:	$-S+(1-\pi)0+\pi R_I \stackrel{!}{=} 0$	$\Leftrightarrow R_I = \frac{S}{\pi}$
JL:	$-2S+(1-\pi)^2 0+(2\cdot\pi(1-\pi)+\pi^2)2R_J \stackrel{!}{=} 0$	$\Leftrightarrow R_J = \frac{S}{\pi(2-\pi)}$

The lender will only accept contracts with exactly these interest rates! Then, however, he is indifferent between the two. From now on we can therefore focus on such contracts and ignore all the rest. Obviously $R_J \leq R_I$, so that part of the argument in [AM10] is correct: Joint liability lowers[74] nominal interest rates, compared to individual liability.

Proposition 1 (Comparison of nominal interest). **With a creditor that just breaks even, nominal interest under joint liability is lower than that under individual liability.**

6.3.2 The debtors' perspective

Which is the contract preferred by a borrower? To answer this question we can adopt various different evaluation criteria.

Evaluation criterion 1: Expected payoff

To begin with, we can calculate the expected payoff for the borrower in both cases and see if there is a difference.

	Expected payoff
IL:	$\pi(A-R_I) = \pi A - \pi R_I = \pi A - S$
JL:	$\pi(1-\pi)(A-2R_J)+\pi^2(A-R_J) = \pi A - \pi(2-\pi)R_J = \pi A - S$

Expected payoff is equal in both cases, and it must always be equal as it can simply be calculated[75] as 'expected ability to repay' minus 'expected repayment'. The ability to repay does not depend on the contract and expected repayment must always be S such that the bank breaks even.

Proposition 2 (Comparison of expected payoff). **The debtor's expected payoff is the same under joint liability and under individual liability. Hence, a risk neutral borrower is indifferent.**

74 Of course, apart from the edge case where $\pi = 1$.

75 By linearity of expectations.

Evaluation criterion 2: Expected repayment

Maybe an inspection of expected repayment in case of success yields more insight. The lower rate R_J might indicate that the borrower gets a cheaper loan under joint liability. How much does the borrower repay in expectation, given he can repay?

	Expected interest
IL:	$R_I = \frac{S}{\pi}$
JL:	$R_J + (1-\pi) \cdot R_J = (2-\pi)R_J = \frac{S}{\pi}$

If a borrower under joint liability is successful he pays for his own loan and with a certain probability he has to pay for his peer. In expectation the repayment is just as high as under individual liability, it makes no odds.

Proposition 3 (Comparison of expected repayment)**.** Although the nominal interest for joint liability is lower than that for individual liability (proposition 1), expected repayment is the same under both contracts.

Evaluation criterion 3: Risk

If there is no discrepancy in expectations, perhaps there is a difference in risk. So let's check if one of the two probability distributions is a mean preserving spread of the other. In order to do this, we have to subtract the cumulative distribution functions (CDF) of both probability distributions and integrate over the difference (see [RS70, integral conditions (6) and (7)]). Let's briefly give the two CDFs names, F_I^D for IL and F_J^D for JL, where D indicates that we are talking about the debtor's probability distribution of payoff. The region between F_I^D and F_J^D forms two rectangles which are of equal size:

$$\begin{aligned}
&(2R_J - R_I) \cdot \pi(1-\pi) - (R_I - R_J) \cdot \pi^2 \\
&= 2R_J\pi - R_I\pi - 2R_J\pi^2 + R_I\pi^2 - R_I\pi^2 + R_J\pi^2 \\
&= R_J(2\pi - \pi^2) - R_I\pi \\
&= S - S = 0
\end{aligned}$$

We can therefore conclude that

$$\int_0^A (F_J^D - F_I^D)\, d\lambda = 0 \text{ and } \int_0^y (F_J^D - F_I^D)\, d\lambda \geq 0 \; \forall y \in [0,A]$$

which just says that F_J^D is a mean preserving spread (MPS) of F_I^D.

Proposition 4 (Comparison of risk). From the debtor's perspective and in terms of mean preserving spreads, a joint liability contract is riskier than individual liability.

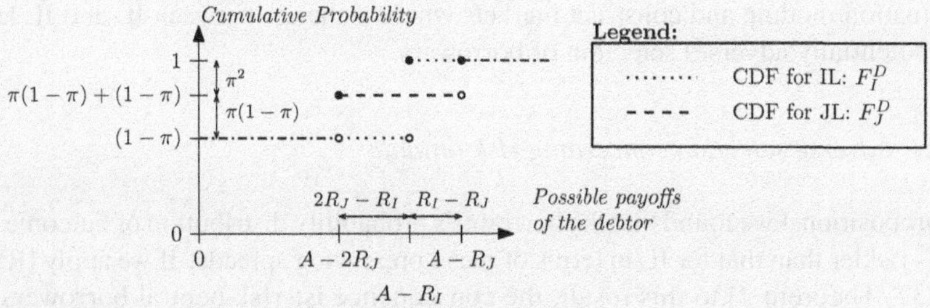

6.4 The debtor's CDF of outcomes under IL and JL
This graph shows the probability distributions of payoff of a borrower in terms of its CDF, for IL and JL. A dashed-dotted line indicates that both CDFs are identical in that interval.

Evaluation criterion 4: Maximal repayment

Finally, let's have a look at yet another possible evaluation criterion, namely maximal repayment. If a borrower with a joint liability contract has to repay his own and his partner's loan, his gross interest is $2R_J$ – compared to R_I under individual liability. A comparison of the two yields:

$$2R_J - R_I = \frac{2S}{\pi(2-\pi)} - \frac{S}{\pi} = \frac{2S - 2S + \pi S}{\pi(2-\pi)} = \frac{S}{2-\pi} > 0\,.$$

Consequently, to be able to afford a joint liability contract, a borrower needs a higher ability to repay A than under individual liability.

Proposition 5 (Comparison of maximal repayment). Maximal repayment under joint liability is higher than under individual liability.

6.4 Adverse selection [contracting module]

Now, can joint liability really mitigate adverse selection? The third part of the argumentation is situated in the contracting module. We take the results from the evaluation module and construct markets where a change between IL and JL leads to (potentially adverse) selection of borrowers.

6.4.1 Adverse selection concerning risk-attitude

In proposition 4 we found that the borrower's probability distribution of outcomes for JL is riskier than that for IL in terms of mean preserving spreads. If we apply [RS70, p. 237, Theorem 2] to this result, the consequence is: risk-neutral borrowers are indifferent between individual and joint liability lending, risk-averse borrowers will prefer IL, and risk-friendly borrowers will gather at the JL contract. With a complete change from IL to JL, risk-averse borrowers may even drop out of the market.

Is this an instance of *adverse* selection, i.e. is it problematic for the lender? Not in this model, as π is publicly known and invariant over time. However, it becomes a problem when ex ante moral hazard is an issue. Risk-friendly borrowers may tend to increased risk taking during the uncertainty phase (section 3.4). Since the interest rate is fixed at the beginning (based on ex ante π), a deterioration of π in the course of the contract reduces the lender's expected repayment who will in turn miss his break-even target. So if a selection effect makes the pool of borrowers risk-friendlier, together with moral hazard this can erode the lender's expected utility and thus constitute a case of adverse selection.

Proposition 6 (Selection effect of JL concerning risk attitude)**.** Consider a market with borrowers that have differing risk-attitudes but are otherwise identical. Then changing from individual liability lending to joint liability lending can push risk-averse borrowers out of the market.

Corollary 1 (Adverse selection of JL concerning risk attitude)**.** In combination with moral hazard the selection effect of JL concerning risk attitude becomes an adverse selection effect, worsening the quality of the pool of borrowers from the lender's perspective.

6.4.2 Adverse selection concerning investment project risk

Let's now consider a different pool of borrowers – the pool that Armendáriz and Morduch assume in their argumentation (see [AM10, p. 102 f.]). It is characterised by the following features:

- There are two types of borrowers. The s-type features a safe investment project, whilst the r-type has access to a risky investment project.
  - Safe investment projects yield A_s with certainty. Risky projects yield A_r with probability $\pi \in]0,1[$ and 0 otherwise.
  - A_s and A_r are sufficiently high to repay even a joint liability contract.
  - Expected returns are the same for both projects, i.e. $\pi A_r = A_s$.
- The fraction of s-type borrowers is $\sigma \in [0,1]$, and the fraction of r-type borrowers is given by $(1-\sigma)$, respectively.
- For joint liability contracts borrowers sort themselves into homogeneous groups of identical types (assortative matching).

Similarly to proposition 1 we can establish

	Break-even condition	Gross interest
IL:	$-S+\sigma[R_I]+(1-\sigma)[\pi R_I] \stackrel{!}{=} 0$	$\Leftrightarrow R_I = \frac{S}{\sigma+(1-\sigma)\pi}$
JL:	$-2S+\sigma[2R_J]+(1-\sigma)[(2\pi(1-\pi)+\pi^2)2R_J] \stackrel{!}{=} 0$	$\Leftrightarrow R_J = \frac{S}{\sigma+(1-\sigma)\pi(2-\pi)}$

If $\sigma = 0$ these are the same results as on page 123. For $\sigma < 1$ we obviously have $R_J < R_I$ again, thus the statement of proposition 1 still holds. Furthermore $R_I < 2R_J$, so proposition 5 is also still true. To see this, calculate

$$2R_J - R_I = \frac{2S(\sigma+(1-\sigma)\pi) - S(\sigma+(1-\sigma)\pi(2-\pi))}{(\sigma+(1-\sigma)\pi)(\sigma+(1-\sigma)\pi(2-\pi))}$$

The common denominator is positive and looking at the numerator we get

$$\begin{aligned} S(2\sigma+2(1-\sigma)\pi-\sigma-(1-\sigma)\pi(2-\pi)) &= S(\sigma+(1-\sigma)\pi[2-(2-\pi)]) \\ &= S(\sigma+(1-\sigma)\pi^2) \end{aligned}$$

which is a positive number. So $2R_J > R_I$, indeed.

By the way, what is the reason for all these results still being true? In section 6.3 we established them for a single contract and an arbitrary Bernoulli distribution. If the comparison is true for every single contract is also true on an aggregate level – independent of the exact fractioning of the pool of borrowers.

Armendáriz and Morduch's argument now seems to be that $A_s < A_r$ (a consequence of the assumption that $\pi A_r = A_s$), so safe borrowers can repay less and may be able to afford the nominally 'cheap' interest rate R_J whilst not being able to afford the nominally 'expensive' contract with R_I. According to [AM10]:

- "The bank can thus reduce the interest rate and lure deserving safe types back into the market." [AM10, p. 104]
- "[...] the bank is able to set interest rates that are low enough to win back the business of the safe borrowers." [AM10, p. 108]

But what does it mean to be able to afford a credit contract? It means to be able to repay the stipulated amount of money. And for a joint liability contract the amount stipulated in case of a defaulting peer is $2R_J$. The bank will not close a joint liability contract with a borrower that cannot repay $2R_J$, i.e. that cannot bail his peer out (it would ruin the fundamental assumption for the calculation of R_J). So, since $R_I < 2R_J$, in fact *less* borrowers can afford the joint liability contract and in particular the safe borrowers with $A_s < A_r$, will drop out of the market first.

Is this a case of *adverse* selection? It is, indeed! Despite the fact that both investment projects have the same expected payoff, borrowers with a risky project repay less often – only in a faction of π of all cases. The inconvenient truth that the creditor is hit negatively by a riskier pool of borrowers is also reflected by the fact that both break-even interest rates increase with a decrease of σ.

$$\frac{\partial R_I}{\partial \sigma} = \frac{-S(1-\pi)}{(\sigma+(1-\sigma)\pi)^2} < 0$$

$$\frac{\partial R_J}{\partial \sigma} = \frac{-S(1-\pi)^2}{(\sigma+(1-\sigma)\pi(2-\pi))^2} < 0$$

Proposition 7 (Selection effect of JL concerning project risk)**.** Consider a market with borrowers whose investment projects differ in their payoff in case of success but who are otherwise identical (i.e. same expected project return, same project return in case of insolvency, same risk-attitude of borrowers, etc.). Then changing from individual liability to joint liability can push the borrowers with lower project payoffs in case of success out of the market.

Corollary 2 (Adverse selection of JL concerning project risk)**.** The remaining pool of borrowers with only risky investment projects deteriorates the lender's expected repayment. The selection effect of joint liability is therefore an instance of adverse selection from the lender's perspective.

So keep in mind that borrowers with low income are squeezed out when changing from IL to JL, since they cannot afford the return service $2R_J$ in the bail-out case of the joint liability contract – that is true in general. A special case appears if we construct the pool of borrowers in our model such that risk and income are inversely related to each other, then we can conclude: save projects are squeezed out. However, this may break down for pools of borrowers with different characteristics, so it isn't always true.

6.4.3 A potential knock-on effect of joint liability and adverse selection

We have just seen two adverse selection effects of joint liability. In real life, with a realistically complex pool of borrowers, the first selection effect may trigger the second one:

1. Under JL risk-averse borrowers drop out of the market (proposition 6).
2. Together with moral hazard expected repayment decreases (corollary 1).
3. In turn the lender has to increase interest rates to maintain break-even.
4. Increased interest rates crowd out lenders with save investment projects (proposition 7).
5. In turn the lender suffers the consequences of a secondary wave of adverse selection (corollary 2) and has to raise interest rates even higher.

The result is a pool of risk-friendly borrowers with risky investment projects.

6.5 Conclusion: Joint liability can causes adverse selection

Which credit contract is better, individual or joint liability? The question is a typical evaluation topic. Therefore, we considered various evaluation criteria to distinguish the two. We reproduced the result of Armendáriz and Morduch that JL goes along with lower nominal break-even interest rates. However, this has no effect on expected payoff for the borrower. Also expected repayment is the same for both contracts. But we have seen that JL contracts increase risk (in terms of mean preserving spreads) for the borrower. Furthermore maximal repayment is higher under JL.

Subsequently, we examined the consequences of these evaluation results on the contracting module of lending theory. Certainly, the selection effects of joint liability potentially exclude borrowers from the market (see proposition 6 and 7) which can lead to an underinvestment problem (welfare perspective). Under certain conditions, however, these selection effects reduce the lender's utility and are therefore *adverse* selection effects (lender's perspective).

Note how the clear separation between the evaluation and the contracting module helped us structure the argumentation. For example, joint liability increases risk for the borrower, that's a fact (evaluation module). But the finding only leads to consequences in markets with not just risk-neutral borrowers (contracting module). It depends on the construction of the pool of borrowers and the market structure how the results from the evaluation module translate to the contracting module.

We have seen no case in which JL can mitigate adverse selection. On the contrary, joint liability is very likely to cause adverse selection with respect to risk-attitude of borrowers and their investment project risk. The claim that joint liability generally mitigates adverse selection should therefore be regarded as proven false. The lower nominal interest rates under JL are misleading. That does not mean that JL contracts are always worse than IL. Since the peer's repayment in a JL setting is not a mere random variable but the result of the actions of a human being, the JL reciprocity agreement can induce secondary effects in the enforcement modules like leveraged peer pressure or induced peer monitoring and thereby improve information and incentive aspects. It is not the reciprocity agreement or the superior repayment function that makes JL interesting, it rather is its potential to trigger and exploit the informational and incentive capabilities of borrowers. If we neglect the enforcement modules, however, joint liability does not improve anything. And of course, beneficial secondary effects must first of all outweigh the disadvantages stemming from the repayment function (e.g. higher maximal repayment, increased risk) that have been highlighted in this chapter.

7 Collateralisation shifts risk towards borrower – A formal proof

Abstract This article formally proves that if, in a standard debt contract, collateral is increased and interest is decreased simultaneously in a way that keeps expected repayment constant, then for the borrower the resulting probability distribution of payoff is a mean preserving spread of the original one. Together with a theorem from [RS70] the consequence is that risk-averse borrowers prefer loans with little collateral. The result has numerous implications on the evaluation and the contracting module of lending theory.

7.1 Collateral and credit risk in the literature

The theory of credit contracts in conjunction with risk have been studied from various perspectives in the literature. Two important authors are the following.

Stiglitz and Weiss wrote a famous paper about credit rationing in markets with imperfect information. They take a pool of many borrowers with different abilities to repay (section 3.3), all with equal mean but differing riskiness.[76] Then they consider a certain credit contract and determine which borrowers will actually borrow. It turns out to be the borrowers with abilities to repay that are riskier than a certain threshold (see [SW81, Theorem 1]).[77] Borrowers with a risky ability to repay seem more affine to borrowing.[78]

In [Bes85] Bester refines the result of Stiglitz and Weiss. In his paper the borrowers not only have the choice between borrowing and not borrowing but can decide between different contracts. The standard credit contracts offered all conform in expected repayment but exhibit different collateral requirements. So instead of

76 They are mean preserving spreads of each other, see [SW81, p. 395].

77 Their paper of course continues with proving that the threshold rises with increasing interest rates, with calculating expected repayment, with determining the optimal interest rate for the lender by contrasting the advantages from higher return services with the drawbacks from a riskier pool of borrowers, and with showing how credit rationing comes about under hidden characteristics – and even more.

78 Likewise, borrowers with two projects of differing risk prefer to play out the riskier one (see [SW81, Theorem 7]). Also see p. 46 for an explanation of the fundamentally conflicting interests between creditors and debtors.

C. Prem, *The Theory of Credit Contracts*,
https://doi.org/10.1007/978-3-658-29362-8_7

studying interest and collateral separately, as in [SW81], Bester lets the two vary simultaneously to demonstrate an instance of self-selection (p. 76). He calculates the interest-collateral indifference curves, showing that borrowers with a risky ability to repay have a steeper one. Given the borrowers' trade-off between interest rate and collateral, together with the zero-profit constraint of the bank, borrowers with a riskier ability to repay end up with contracts with low collateralisation.

In a related paper [Bes87] Bester again discusses the role of collateral. This time with "an arbitrary number of investors who are taken to be risk-averse" [Bes87, p. 888]. Note that all borrowers share the same utility function, in particular the same degree of risk-aversion. Furthermore, in [Bes87] collateralization "is limited by the borrowers' initial wealth." [Bes87, p. 888]. Again, borrowers with risky investment projects end up with little collateral (see [Bes87, p. 893]).

All these models are monolithic lending models (chapter 2). They mix reciprocity and evaluation topics with contracting topics like adverse selection, rationing and self-selection without establishing a satisfying degree of generality in either of these modules. With specific assumptions about the pool of borrowers, their wealth, their evaluation criteria, their ability to repay, the market structure, etc. they are very special showcases with a narrow scope.

In contrast, this article will focus and generalise. We will prove in general how the probability distribution of payoff changes with different degrees of collateralisation. The entire analysis is situated on the reciprocity layer of lending theory (p. 16). Only then we discuss possible implications on related modules. Separating these two research steps delivers general results[79] and the interdependencies between different modules become crystal clear.

7.2 Concepts: Risk, collateral, standard debt contracts

The term 'risk' can be located in two places on the map of lending theory (p. 16). On the one hand it can appear in conjunction with the random variable 'ability to repay'. On the other hand it can be found in the evaluation module when describing the risk-attitude of debtors or creditors. It is important to keep these two issues clearly separated. A risk-averse borrower may be the owner of a relatively safe investment project and vice versa. So speaking of a 'risky borrower' is certainly ambiguous. To compare the riskiness of several random variables with common mean we use the concept of mean preserving spreads (MPS) presented in [RS70].

79 The proposition will be independent of a certain type of probability distributions, a certain risk-attitude, a certain construction of the pool of borrowers, etc.

There are different ways to model collateral (p. 42). In this article the term 'collateral' will be used to refer to non-stochastic outside collateral. That is the ability to repay is composed of a random variable plus a non-stochastic constant. To have an unambiguous way of speaking, the former will be called the *stochastic ability part* to distinguish it from the non-stochastic component. Both, creditor and debtor, agree on their judgement of these resources (no asymmetric valuation). Since we want to examine the difference between various degrees of collateralisation we assume that the borrower is in a situation where he can accede to contracts from zero to full collateralisation.

In this article we compare different members of the family of standard debt contracts. Formally, standard debt contracts are contracts with a repayment function from the following set:

$$\{r\colon \mathbb{R}_{\geq 0} \to \mathbb{R}_{\geq 0}, a \mapsto r(a) \mid r(a) = \min\{a+C, R\}, 0 \leq C \leq R\}$$

Here, a stands for the realisations of the stochastic part of the ability to repay, and R and C are two components of the stipulated return service, i.e. gross interest (principal plus interest) and collateral, respectively.

7.3 Calculating the transformed distribution

Before we start with the actual proof of the main proposition in this article we need some preparation. In order to apply the integral conditions for a mean preserving spread (see [RS70, Eq.(6),(7)]) we need to calculate the cumulative probability distribution (CDF) of a borrower's final payoff (see Lemma 1).

Notice that the size of the loan is not considered here. To incorporate it, shift the creditor's distribution to the left according to the size of the loan. Loan size would be necessary to calculate the lender's *exact* expected payoff. We only need expected payoff to be constant – no matter at which magnitude – and thus loan size is irrelevant for the following argumentation.

Notation.

$a \in [0, \bar{A}]$	Possible realisations of the ability to repay
$F\colon \mathbb{R} \to [0,1]$	CDF of the ability to repay
$R \in [0, \bar{A}]$	Gross interest
$C \in [0, R]$	Collateral
$t\colon [0, \bar{A}] \to \mathbb{R},\ a \mapsto \max\{-C, a-R\}$	Borrower's payoff in each state
$F^t\colon \mathbb{R} \to [0,1]$	CDF of the borrower's prob. distr. of payoffs

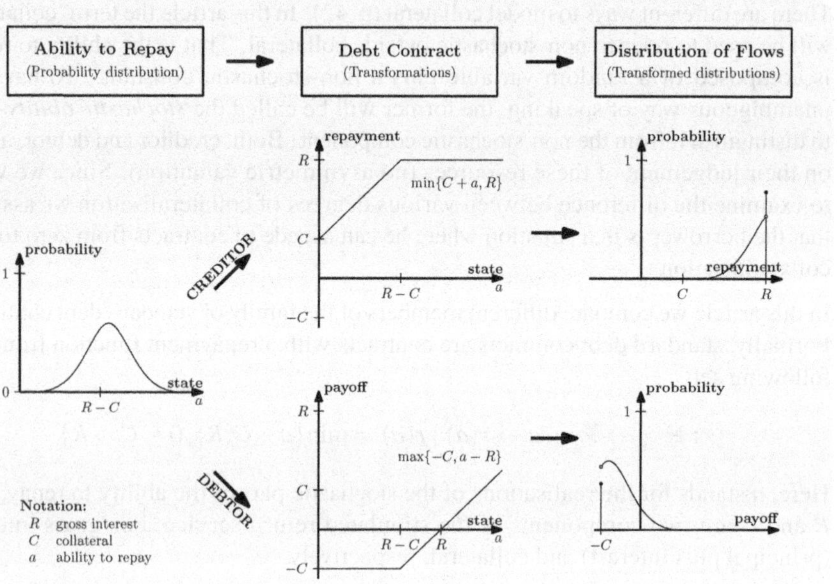

7.1 **Transformations in a standard debt contract**
The figure shows how a standard debt contract leads to the creditor's and debtor's payoff distributions. From a probability theory perspective it is a transformation that shifts and truncates the ability to repay. The result are mixed distributions with continuous and discrete parts.

Lemma 1. The borrower's transformed distribution F^t is

$$F^t(x) = \begin{cases} 0 & x < -C \\ F(R+x) & x \geq -C \end{cases}$$

Proof. According to the definition of a cumulative probability function (CDF)

$$F^t(x) := \Pr(t(a) \leq x) = \Pr(t^{-1}]-\infty, x]) .$$

So we have to calculate the probability of the preimage of $]-\infty, x]$ under t.

Let's first of all explicate the transformation t.

$$t(a) := \max\{-C, a-R\} = \begin{cases} -C & a < R-C \\ a-R & a \geq R-C \end{cases}$$

Then the preimage of the set $]-\infty,x]$ is

$$t^{-1}(]-\infty,x]) = \begin{cases} \emptyset & x < -C \\ [0,R+x] & x \geq -C \end{cases}$$

The first case is obvious, since t never takes values below $-C$. The preimage of $x = -C$ is given by all values of a that lie below $R - C$, i.e. by the interval $[0,R-C]$. And the preimage for any other $x > -C$ is simply the same as the preimage of the line $a \mapsto a - R$.

Using F we can express the probability

$$\Pr(t^{-1}(]-\infty,x])) = \begin{cases} 0 & x < -C \\ F(R+x) & x \geq -C \end{cases}$$ □

7.4 The risk-shift of collateralization

Now we want to compare two standard debt contracts, one with low collateral C_l and another one with high collateral C_h. The ability to repay and its cumulative distribution function $F\colon \mathbb{R} \to [0,1]$ remain the same, but we will get two different transformed distributions F^l and F^h. Therefore we enrich our notation with the indices l and h.

	Contract l with low collateral	**Contract h with high collateral**
Collateral:	C_l	C_h
Gross interest:	R_l	R_h
CDF of borrower's payoff:	F^l	F^h

Remark 1. If we calculate $F^l(x)$ and $F^h(x)$ according to Lemma 1, we recognise that left of $-C_h$ we have

$$F^l(x) = 0 = F^h(x) \; \forall x < -C_h$$

and right of $\bar{A}$ we have

$$F^l(x) = 1 = F^h(x) \; \forall x > \bar{A}\,.$$

So in both regions the transformed CDFs coincide and thus their difference is zero. This remark says that it is sufficient to integrate their difference over compact intervals and avoid improper integrals.

Remark 2. Also notice that for any combination (R,C) we have

$$\min\{a+C,R\}+\max\{-C,a-R\}=a$$

which says that the repayment plus the borrower's payoff in every state adds up to the realisation of the ability to repay. This almost trivial statement can be proven by a simple distinction of cases, $a \gtrless R-C$, and can be observed in figure 7.1. What is important for us is that – by virtue of the linearity of expectation – the equation also holds in terms of expectations of these random variables. So expected repayment plus expected borrower payoff equals $\mathbb{E}(a)$ for any contract (R,C).

To prove our main proposition we need the following technical lemma.

Lemma 2. Take two standard debt contracts (R_l,C_l) and (R_h,C_h) with $C_l < C_h$ such that both contracts generate the same expected repayment – given the ability to repay represented by $F\colon \mathbb{R} \to [0,1]$. Then:

1. The borrower's expected payoff is the same under both contracts.
2. $\int_{-C_h}^{-C_l} F(R_h+x)\,dx = -\int_{-C_l}^{\bar{A}} (F(R_h+x)-F(R_l+x))\,dx.$
3. $R_h \leq R_l$.

Proof. To prove claim number one, notice that if expected repayment in case of contract (R_l,C_l) is the same as for (R_h,C_h), then remark 2 tells us that also expected payoff for the borrower must be equal for both contracts.

This fact can be expressed in terms of CDFs by using the integal condition

$$\int (F^h(x)-F^l(x))\,dx \stackrel{!}{=} 0\,.$$

Since the integrand is zero outside the compact interval $[-C_h,\bar{A}]$ (see remark 1) this is a proper integral. It can be split into

$$\int_{-C_h}^{-C_l} (F^h(x)-F^l(x))\,dx + \int_{-C_l}^{\bar{A}} (F^h(x)-F^l(x))\,dx = 0\,.$$

Using lemma 1 to calculate F^h and F^l this becomes

$$\int_{-C_h}^{-C_l} F(R_h+x)\,dx + \int_{-C_l}^{\bar{A}} (F(R_h+x) - F(R_l+x))\,dx = 0$$

which proves the second claim.

Finally, to see $R_h \leq R_l$, notice that the first addend in the last sum is certainly non-negative. (If $F(R_h - C_h) > 0$, it is positive.) Thus, for the sum to be zero, the second integral must be negative. In turn the integrand can only be negative, that is $F(R_h+x) \leq F(R_l+x)$, for at least some x. But both terms are just translations of the same function, so actually it will be $F(R_h+x) \leq F(R_l+x)$ for either none or all x. Since F is a CDF and hence increasing, the inequality of values implies the following inequality of arguments $R_h + x \leq R_l + x$ or simply $R_h \leq R_l$. (If $F(R_h - C_h) > 0$, the inequality is strict.) □

Having established Lemma 2, we are in the position to prove the central result.

Proposition 1 (Risk–Shift of collateralisation). Consider two standard debt contracts (R_l, C_l) and (R_h, C_h) with $C_l < C_h$ such that both contracts generate the same expected repayment – given the ability to repay represented by the CDF $F: \mathbb{R} \to [0,1]$. Let F^l and F^h denote the debtor's probability distribution of payoffs under each contract, respectively. Then F^h is a mean-preserving spread of F^l.

Proof. By assumption and Lemma 2 (claim 1) both F^h and F^l have the same mean. What's left to show is the integral condition

$$\int_{-\infty}^{z} (F^h(x) - F^l(x))\,dx \geq 0 \;\forall z.$$

Since the integrand is zero outside of $[-C_h, \bar{A}]$ (see remark 1) the integral is zero for all $z \leq -C_h$ and if z rises above $\bar{A}$ the value of the intergal will not change anymore. So only integration limits $z \in [-C_h, \bar{A}]$ remain relevant to prove the claim.

If $-C_h \leq z \leq -C_l$ we can use Lemma 1 to write the integral as

$$\int_{-C_h}^{z} (F^h(x) - F^l(x))\,dx = \int_{-C_h}^{z} F(R_h+x)\,dx.$$

which is clearly non-negative. (It is positive if $F(R_h - C_h) > 0$.)

In case $-C_l < z < \bar{A}$ we again use Lemma 1 to write the integral in the following form

$$\int_{-C_h}^{z} (F^h(x) - F^l(x))\,dx = \int_{-C_h}^{-C_l} F(R_h + x)\,dx + \int_{-C_l}^{z} (F(R_h + x) - F(R_l + x))\,dx\,.$$

Now the first integral is ≥ 0, the second one is ≤ 0. The challenge is to determine the sign of the sum. Apply Lemma 2 (claim 2) to replace the first addend and get

$$-\int_{-C_l}^{\bar{A}} (F(R_h + x) - F(R_l + x))\,dx + \int_{-C_l}^{z} (F(R_h + x) - F(R_l + x))\,dx\,.$$

The two integrals have the same integrand, thus can be combined and yield

$$\int_{z}^{\bar{A}} (F(R_l + x) - F(R_h + x))\,dx$$

which is non-negative because $R_l \geq R_h$ (lemma 2, claim 3) and since the CDF F is increasing, like all CDFs. (It is positive if $R_l > R_h$ and $F(R_h - z) < 1$.) □

The proposition says that, everything else equal, more collateral makes the standard debt contract riskier for the borrower. What is well in line with our intuition has now been formally proven. If we emerge from the details and look back, the main steps have been to calculate the transformed distributions F^l and F^h and then prove the integral conditions for mean-preserving spreads.
It is only natural to ask if – in return – high collateral reduces the risk for the lender. To do this we enrich our notation a second time to distinguish lender and borrower. Let $F^l_{\mathcal{C}}$ and $F^l_{\mathcal{D}}$ denote the transformed distributions of the creditor and the lender, respectively.

Proposition 2 (Risk-shift of collateralisation)**.** Take two standard debt contracts (R_l, C_l) and (R_h, C_h) with $C_l < C_h$ such that both contracts generate the same expected repayment – given the ability to repay represented by the CDF $F : \mathbb{R} \to [0,1]$. Let $F^l_{\mathcal{C}}$ and $F^h_{\mathcal{C}}$ denote the creditor's probability distribution of repayment under each contract, respectively. Then $F^l_{\mathcal{C}}$ is a mean-preserving spread of $F^h_{\mathcal{C}}$.

Proof. The proof is similar to the argumentation above, so I will just sketch it. Calculate the transformed CDF $F^t_{\mathcal{C}}$ for an arbitrary contract (R,C).

$$F^t_{\mathcal{C}}(x) = \begin{cases} F(-C+x) & x < R \\ 1 & x \geq R \end{cases}$$

Recycle the fact that $R_h \leq R_l$ from Lemma 2 (claim 3) and calculate

$$F^l_{\mathcal{C}}(x) - F^h_{\mathcal{C}}(x) = \begin{cases} F(-C_l+x) - F(-C_h+x) & x < R_h \\ F(-C_l+x) - 1 & R_h \leq x < R_l \\ 1-1 & R_l \leq x \end{cases}$$

In particular the difference is zero outside the interval $[0,R_l]$, since $a \in [0,\bar{A}]$. The challenge again is to determine the sign of the second integral condition if $z \geq R_h$.

$$\begin{aligned} \int_{-\infty}^{z} (F^l_{\mathcal{C}}(x) - F^h_{\mathcal{C}}(x))\,dx = & \int_0^{R_h} (F(-C_l+x) - F(-C_h+x))\,dx \\ & + \int_{R_h}^{z} (F(-C_l+x) - 1)\,dx \\ = & \ldots \end{aligned}$$

and exploiting the first integral condition (same mean) to replace the first addend

$$\begin{aligned} \ldots &= -\int_{R_h}^{R_l} (F(-C_l+x) - 1)\,dx + \int_{R_h}^{z} (F(-C_l+x) - 1)\,dx \\ &= \int_z^{R_l} (1 - F(-C_l+x))\,dx \geq 0. \end{aligned}$$

□

The geometric intuition behind both risk-shift results is that values of C that are closer to R reduce the concavity of the repayment function and the convexity of the borrower's payoff (see figure 7.1). For the extreme case of $C = R$ the repayment function is a constant and the borrower's payoff is a straight line.

Just for curiosity, we can ask how all the CDFs involved, that is $F^t_{\mathcal{C}}, F^t_{\mathcal{D}}$ and the original F, are connected to each other. Each of the two transformed distributions holds only part of the information of the original CDF. But how do we have to attach the two in order to regain the distribution of the ability to repay?

Remark 3. An adequate shift of the two transformed distributions yields

$$F^{\prime}_{\mathcal{D}}(x-R) = \begin{cases} 0 & x < R-C \\ F(x) & x \geq R-C \end{cases}$$

$$F^{\prime}_{\mathcal{C}}(x+C) = \begin{cases} F(x) & x < R-C \\ 1 & x \geq R-C \end{cases}$$

thus we can reconstruct all the information about F using

$$F(x) = F^{\prime}_{\mathcal{D}}(x-R) + F^{\prime}_{\mathcal{C}}(x+C) - \mathbb{1}_{[R-C,\infty[}(x) \,.$$

Here $\mathbb{1}_{[R-C,\infty[}$ is the indicator function of the set $[R-C,\infty[$.

7.5 Implications on other lending modules

The argumentation above is situated completely on the reciprocity layer of lending theory. Neither did we use any assumptions about the creditor's and debtor's utility functions or use any concrete measures of risk aversion (e.g. HARA, CARA, IARA, DARA, CRRA, IRRA, DRRA) nor did we revert to any market structure. We did not even use any specific assumptions about the stochastic part of the ability to repay.[80] Its small set of assumptions makes our proposition a very general result. We have established a claim that holds true independently of the concrete configuration of other modules of lending theory. Nevertheless, let's discuss how our result is connected to neighbouring modules.

Reciprocity layer

Although lemma 2 is mainly technical, it has an interesting feature. Its third claim shows that higher values of collateral are associated with lower values of gross interest.[81] Why is that interesting? For comparison, consider the usual argumentation: High collateral is associated with low interest rates because collateral acts as a signal, indicating a borrower with a safer stochastic ability part, and therefore allowing for a

80 Beside the reasonable restriction that it is bounded, see p. 133.

81 An alternative way to show this would be to explicitly calculate expected repayment for an arbitrary standard debt contract (R,C) (lemma 2 only uses the fact that expected repayment is equal, no mater at what magnitude) and to establish the downward-sloping R-C indifference curve by using the implicit function theorem. However, this alternative proof would lose generality as it required additional assumptions on the existence of various partial derivatives.

lower risk premium, i.e. a lower interest spread[82] (see [GW11, p. 51]). According to lemma 2, the detour using hidden characteristics is not necessary at all. We did not need a signalling game to establish this relationship. The stochastic ability part did not even change. We have formally proven that the negative relationship between R and C is just a matter of keeping expected repayment constant for different amounts of collateral.

Evaluation module

By using [RS70, p. 237, Theorem 2] we can jump to the evaluation module and transform proposition 1 into a statement connected to the risk aversion of debtors: Risk-friendly borrowers prefer SDCs with more collateral (and lower interest); risk-averse borrowers choose less collateral (and higher interest).

This implication is particularly important if we reconsider the papers [Bes85] and [Bes87] mentioned in the beginning. Recall that in both papers all borrowers share *the same risk attitude*. Imagine the following variation of Bester's model. Instead of borrowers with different stochastic ability parts we take a pool of borrowers with the same stochastic ability but *different risk attitudes*. What would happen when offering two contracts on the indifference curve of the bank (i.e. same expected repayment) with differing collateralisation? The risk-friendly borrowers would gather at the contract with high collateral.[83]

Altogether, if we let the riskiness of the ability to repay *and* the risk-attitude of borrowers vary, two effects mix: the self-selection effect concerning the riskiness of the ability to repay (presented by Bester) and the self-selection effect concerning the risk-attitude of borrowers (presented here). This may explain why Bester's result, where borrowers with risky investment projects end up with little collateral ([Bes85, p. 854], [Bes87, p. 893]), is rarely observed in reality (see [BU90, MNS06]). It may be blurred by the risk-attitude selection effect. A borrower with a rather safe ability to repay (who would pledge high collateral according to Bester) may choose the contract with little collateral if he is extremely risk averse.

---

82 The difference between borrowing and lending rate.

83 Given a model with a borrower decision point that makes ex ante moral hazard an issue, risk-friendly borrowers can be just as problematic to the lender as borrowers with risky stochastic ability parts. They may engage in more risky investment decisions, which is detrimental to the bank (see [SW81, Theorem 3]). Different risk attitudes, collateralisation and ex ante moral hazard are brought together in the paper [Coc99]. Instead of borrowers with different but exogenously fixed risk-attitudes and different but exogenously fixed stochastic ability parts Coco integrates the possibility of ex ante moral hazard by making the stochastic ability parts endogenous. He lets borrowers choose amongst several different investment projects – according to their risk attitude.

Contracting module

When jumping into the contracting module of lending theory the results of our propositions can produce statements about the bargaining power of each party.

For instance, imagine a risk-averse lender (p. 69) that is confronted with a risk-averse pool of borrowers. Given both parties agree on the expected repayment of the credit contract, creditor and debtor can still negotiate how the risk is shared. Now it depends on the bargaining power of each party whether the contract will exhibit high collateralisation or not. Borrowers with a weaker bargaining position will end up with contracts with higher values of collateral.

Altogether: If risk distribution is negotiable matter and risk-attitudes are unidirectional, then the collateralisation of a loan may tell a story about the relative bargaining power of creditor and debtor.

7.6 Summary and further research

Summary

Let's recap the milestones in this article. Technically, we wrote down and illustrated the flows of value in a standard debt contract (SDC) similarly to [SW81, Figure 2]. In addition, however, we examined how these functions transform the probability distribution of the ability to repay and we worked out the transformed cumulative distribution functions (CDFs). Finally, we calculated the integral conditions to formally prove that collateralisation is responsible for a risk-shift from lender to borrower.

Subsequently, we investigated the implications of this result. Consequent adherence to the modular research philosophy helped to disentangle the various interrelationships. On the reciprocity layer of lending theory we found a negative relationship between interest and collateral without any recourse to different abilities to repay or asymmetric information. In the evaluation module we found that the peference of a borrower for collateralisation is dependend on his risk-attitude; risk-averse borrowers avoid collateral. Eventually, in the contracting module even a risk-averse borrower may end up with high collateral if he faces a risk-averse lender with a superior bargaining position.

Consequences for empiric research

Why do banks request collateral? The literature typically provides two – seemingly contradictory – arguments to rationalise the frequent use of collateralised debt contracts (see [GW11]):[84]

Observed-risk hypothesis The lender can observe the distribution of the ability to repay and uses collateral to limit maximal loss. This hypothesis predicts a positive correlation between credit risk and collateral.

Adverse selection hypothesis The lender cannot observe the distribution of the ability to repay but the borrowers use collateral to signal their soundness. This hypothesis predicts a negative correlation between credit risk and collateral.

Both hypotheses centre on the question of how the riskiness of stochastic ability parts are related to collateral.

Empiricists have tried to validate one of these hypotheses and disqualify the other.[85] According to Godlewski and Weill "empirical evidence tends to support the observed-risk hypothesis, according to which riskier borrowers are more often required to provide collateral." [GW11, p. 53]. They say that the contradiction can partly be resolved by incorporating the degree of hidden characteristics. Godlewski and Weill find that "a lower degree of information asymmetries enhances the positive relation between the presence of collateral and the loan spread" [GW11, p. 76].

Now, note that the first quotation speaks of the connection between collateral and "riskier borrowers"; the second talks about the relation between collateral and "loan spread". But that's two different kettle of fish! In general, using R as a correlate for credit risk is not very convincing, as the interest spread is more than a mere risk premium (see p. 34). Credit ratings from rating agencies or scores from consumer reporting agencies would be a better measure of the lender's ex ante beliefs about the borrower's ability to repay. Our results prove that finding a negative correlation between R and C in the data can simply be based on the fact that the lender wants to fix (ex ante believed) expected repayment on a certain level. It may have nothing to do with differing stochastic ability parts at all (see lemma 2). Take away: Such R-C-correlations are useless to verify the above hypotheses.

84 For more theory on the use of collateral, consult [Coc00].
85 For an overview see [GW11, Section 2.1].

Further theoretic research

From empirics, let's turn to theory and suggestions for further research. Niinimäki discovered that the indifference-curves in [Bes85] change if collateral is modelled stochastically and both random variables are positively correlated (see [Nii11, p. 2785]). This finding of course also impacts our lemma 2. So future research should research the risk-shift of collateralisation under circumstances where the value of collateral is stochastic.

Justifying the subtitle of this thesis, let's throw a glance at group lending. Recall p. 124, where we came across a similar risk-shift which was induced by joint liability. But it was observed under rather primitive assumptions: There has been only full joint liability instead of different degrees of liability, the abilities to repay were discrete Bernoulli distributions and the dependence structure was the independence copula. A nice challenge would be to establish the risk-shift proposition for joint liability in full generality. Preliminary work has already been performed: A continuous version of a (partial) joint liability repayment function has been presented on p. 39. Calculating the preimages seems manageable. Expressing the probability of these preimages in terms of the original joint probability distribution may be a challenge and require integration over more complicated geometries like ❸ and ❹ in figure 3.8. Establishing the integral conditions may be difficult. Overall, the whole result may heavily depend on the copula. If it can be proven, the implications of such result in the evaluation model should be straightforward, but the implications on the contracting module must be rethought since contracting is a more complicated matter in the case of group lending (see section 3.7.4).

Even more theoretic questions arise when comparing collateral and joint liability:

- Is stochastic collateral in some form similar to joint liability as in both cases the value of the security is uncertain?
- What does the J-C-indifference-curve look like, i.e. how much collateral is equivalent to how much joint liability?
- Does the choice between J and C offer interesting screening mechanisms?

In the microfinance literature the assumption typically is that borrowers have no collateral as they are too poor and the comparison is made between joint liability and individual liability without collateral. But wealthier borrowers could theoretically choose between joint liability and collateral. What are the advantages of each of the two lending tools? A detailed comparison may be insightful – especially with the recently observed shift of microfinance institutions from joint liability to individual liability in mind.

8 Meaning of moral hazard is ambiguous – An improved definition

Abstract This article reflects on the notion 'moral hazard', a concept that refers to some undesired action of the agent in a principal agent setting. Building on the elaborate investigation of [RC12] and after an analysis of previous attempts to define moral hazard, a new definition is proposed. Subsequently, this general definition is applied to credit contract theory.

8.1 The necessity of well-defined concepts

In research concepts evolve and get refined over time. For example the development of the beautifully general and precise modern definition of the limit of a sequence has probably started with the paradoxes that involve limiting processes formulated by the Greek philosopher Zeno of Elea, continued with Newton in his "Analysis with infinite series", until Bolzano and Weierstrass – in the 19th century – formulated the modern definition. Likewise, the concept of moral hazard has evolved over time and the present article is just another effort to advance science by improving the quality of its definitions.

The challenge in precisely defining abstract concepts is to develop formulations that avoid even the vaguest notion of ambiguity but still capture the concept with a certain degree of generality. A single example could certainly be ambiguity free but would be unsatisfying due to its lack of generality. Therefore, defining what is meant by 'moral hazard' is not a trivial task.

Nevertheless the effort that goes into developing precise definitions is necessary, as working with vague terms is confusing and hinders collaboration like in the famous story about the Tower of Babel. Being sloppy about concepts will do anything but result in professional science.

C. Prem, *The Theory of Credit Contracts*,
https://doi.org/10.1007/978-3-658-29362-8_8

8.2 Definitions in the literature and its shortcomings

Roughly speaking moral hazard refers to some unwanted opportunistic behaviour of the agent in a principal-agent-setting. But making this intuition precise is a challenge. Let's have a look at some authors' definitions.

If you research the term 'moral hazard' you most likely stumble across a whole chapter devoted to the topic in the "Handbook of Game Theory" edited by Aumann and Hart. They present the example of an owner of an enterprise that "cannot directly monitor the agent's action nor can he costlessly observe all relevant aspects of the environment." [DR94, p. 870] At the end of the example the authors simply state: "This is the well-known "principal-agent" problem with moral hazard." [DR94, p. 870] Of course, defining concepts by example is not a very satisfying fashion.

Armendáriz and Morduch write: "Moral hazard in lending refers to situations where the bank's risk is tied to unobservable choices made by borrowers." [AM10, p. 48] This definition mentions risk as the thing at stake and specifically refers to hidden actions. In that point it coincides with the previous explanation. But strategic default, for example, is not a *hidden* action, nor is it about risk sharing. Yet, we would call it a moral hazard. Unfortunately, this definition excludes situations that should fall under the term of moral hazard.

According to Mas-Colell, Whinston and Green informational problems arise from hidden information and hidden actions and the hidden action case is "known as moral hazard" [MCWG95, p. 477]. Again, strategic default would not be covered by this definition of moral hazard.

Likewise for the following definition: "Moral hazard refers to the problem of inducing agents to supply proper amounts of productive inputs when their actions cannot be observed and contracted for directly." [Hol82, p. 324]

Eventually, a more general attempt to define moral hazard is provided by Mark William Jenkins: "Most generally, these [moral hazards] include all actions undertaken by the borrower that would not have been undertaken if the borrower's incentives were perfectly aligned with those of the lender." [Jen09, p. 90] The problem with this definition are pathologies like the following: Imagine a MFI (microfinance institution) which is financed by subsidies and donations and therefore has no incentive to put much effort in holding their repayment rates up high. Does strategic default stop being a moral hazard because the lender itself has no incentive to stimulate repayment?

Even more examples for different definitions of moral hazard can be found in Rowell's and Connelly's interesting text: A History of the Term "Moral Hazard". [RC12] They stress the difference between normative and positive definitions of moral hazard. In economics we definitely need a positive definition to work with. But unfortunately the paper does not contain any satisfactory version.

What we have learned so far is that many authors understand moral hazard as 'a situation' that arises in the presence of hidden actions. In contrast, [Jen09] refers to parts of an action set and uses the lender's incentives to define which actions are considered as moral hazard. The difference: Some authors try to define the malady 'moral hazard' via its causes (e.g. asymmetric information), Jenkins uses its symptoms instead (e.g. some bad behaviour).

8.3 A new definition that references the contract

Why is defining the term moral hazard so difficult? Let's proceed step by step. The basic ingredient of a definition of moral hazard is the agent's action set. Some of these actions shall be labelled as moral hazard, as 'bad' behaviour. The difficulty lies in finding a precise positive specification of the normative term 'bad' behaviour. Defining 'bad' relative to the principal's incentives has its problems (see above); in general, defining 'bad' relative to the principal's interests is problematic (e.g. a lender would of course be interested in even more than the stipulated repayment).

A possible solution seems to be to define 'bad' relative to the contract. For instance consider a borrower that does not bail out his peer. Under individual liability this is perfectly fine, whilst with a joint liability contract we would call it moral hazard. Therefore I suggest the following definition.

Definition (Moral hazard). Moral hazard refers to all actions of an agent that intentionally deviate, to the detriment of the principal, from explicitly or implicitly stipulated contract terms.

Most importantly note that the definition's tipping point are the contract terms. Consequently, which actions fall in the moral hazard category changes with the contract.[86] The word 'intentionally' is important to separate e.g. non-strategic default due to bankruptcy from strategic default. It indicates that the agent has had a better choice. Including 'implicit contract terms' is necessary to include e.g. actions of the borrower that increase the risk for the bank even in cases where the utilisation

86 Moreover, it can even change within the contract, see p. 53. This is due to the 'intentionally' aspect of the definition of moral hazard.

of the loan is not explicitly written down. It is an attempt to consider the incomplete contracts branch of contract theory. And 'to the detriment of the principal' specifies the direction of the deviation (favourable deviations from the contract would not be called 'bad' behaviour).[87]

The above definition also draws a clear line between the problems of moral hazard and adverse selection. Adverse selection does not fall under the definition of moral hazard as deviations from a contract are meaningless as long as the contract is not made.[88]

Of course, the definition proposed above still has its weakness. Its reference to implicit contract terms, which are hard to capture, is a soft spot. When leaving it out we miss certain instances of moral hazard, especially if the contract is not formulated comprehensively and precisely enough. But adding it to the definition makes it slightly fuzzy. I therefore hope that future research will manage to improve our notion of moral hazard even further.

The definition above intentionally does not refer to any specific origin of moral hazard. But why may it be attractive for the agent to choose from the moral hazard portion of his action set? The underlying causes that render moral hazard favourable are manifold. It can be informational asymmetries like hidden states or hidden actions but it can as well be weak punishment capabilities of the principal. To prevent moral hazard we have to fight these prerequisites with information tools (e.g. see 3.1.1) and incentive tools (e.g. see 3.1.2).[89]

Eventually, figure 8.1 shows how the general definition can now be used to consistently apply the concept of moral hazard to credit contract theory.

87 Note that this depends on the principal's preferences. Also note the difference between incentives and preferences – and therefore another difference to Jenkins' definition. A lender may prefer higher repayment rates but may have no incentive to put much effort in raising repayment rates if the costs to do so are too high. Preferences refer to the preference ordering on the set of outcomes, incentives refer to the relative attractiveness of the options of an action set at a decision point.

88 On the distinction between moral hazard versus adverse selection and its importance, see [RC12, p. 1059, 1069, 1072].

89 That is why the enforcement triangles consist of the two halves 🖂 and ⚡, see p. 24.

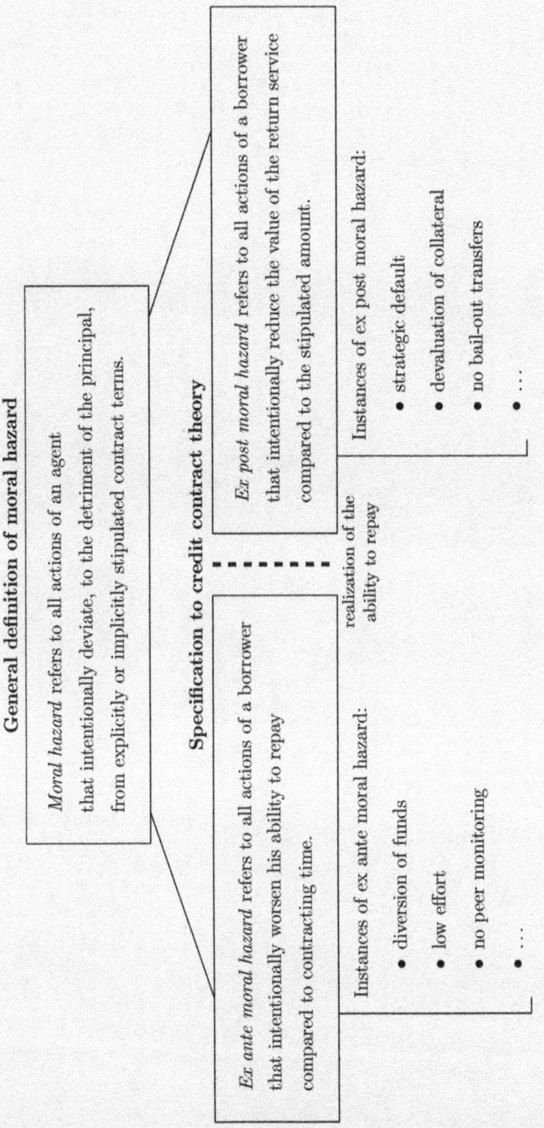

8.1 The definition of moral hazard specified to credit contract theory
This figure subsumes the definitions of moral hazard in credit contract theory under the general definition. Due to the uncertainty involved moral hazard dissociates into ex ante and ex post moral hazard, i.e. into actions before and after the realisation of the random variable.

9 An algorithm that can solve repayment games automatically

9.1 Motivation and development

Lending theory is full of research on repayment games. There are games with individual liability, with joint liability, with joint liability and social sanctions, with joint liability and message games, with partial joint liability and an even greater multitude of repayment games can be imagined where repayment is made cooperatively or non-cooperatively, simultaneously or sequentially, with punishment functions that are symmetric or asymmetric, with two or more borrowers, and so on.

With all these ideas in mind I had to find a way to efficiently calculate the equilibria of many repayment games. To not waste hours of research for tracing Nash equilibria, a first idea was to find students that help and calculate the repayment games I specified. However, marking best responses and finding equilibria by hand is boring and error prone, especially in the huge matrices that we encounter in repaxyment games – each shape is a new game, and each game has several action combinations (see p. 61). Since solving repayment games seems to be a rather mechanical task, the second idea was to search for software that could perform it. Indeed there exists software to calculate the equilibria of games, like "Gambit" (http://gambit-project.org/). Unfortunately, it works only with real numbers. Of course, a bijective function could be used to map a symbolic preference order to the real line. But that also felt rather tedious and could end up in confusion. In a third attempt I even considered using Microsoft Excel. I tried to use its conditional formatting capability to mark best responses automatically. Finally, I realised that a neat solution would require writing a new piece of software that can do the task. I decided to try and meet the challenge – maybe slightly underrating the difficulties ahead. Finding the best action in an action set and combining best responses to Nash equilibria seemed so easy that it should be implemented within a few days.

The foremost challenge was to find a suitable programming language. After some research I turned to Mathematica from Wolfram Research due to its capabilities in symbolic computation. So I worked myself into the theory of operation of Mathematica and studied its syntax. The software licence was provided by the University of Regensburg.

C. Prem, *The Theory of Credit Contracts*,
https://doi.org/10.1007/978-3-658-29362-8_9

Simultaneously, I took online courses on www.edx.org. The series is called "How to Code: Systematic Program Design", comes in three parts and is provided by the University of British Columbia.[90]

After several days of preparation the actual work began. That included thinking about the data types that will be required, splitting the whole algorithm into reasonable sub-problems, writing problem statements, and of course, eventually writing and testing the code. Concerning the problem statement I found it important to write an algorithm that does not only return equilibria but also reveals in a traceable way how it calculated them. It ought to deliver the result as well as its proof such that a human being can confirm it.

Altogether it took me several weeks to develop that program which can solve repayment games automatically. One might object that spending a few days solving repayment games on paper would have been faster. Certainly, it is a huge initial investment. The benefits, however, are numerous – and even more than I expected. The number one benefit is of course the ability to solve games quickly. This ability tremendously increases the motivation to invent and try many different examples of repayment games. Secondly, given the stupidity of the task, a computer seems more reliable than a bleary-eyed PhD student. Thirdly, repayment games with more than two or three borrowers can hardly be solved by hand. Imagine three borrowers that can repay either $0, 1R, 2R, 3R$. That means the unrestricted shape alone has $4 \cdot 4 \cdot 4 = 64$ entries – but there are 64 shapes. A fourth and unexpected gain turned out to be the necessity to gain an absolutely perfect understanding of the formal structure of repayment games to be able to write this software. Even though you thought you got it, having to explain something to a computer tremendously sharpens your comprehension of the problem. The flows-of-value concept stemmed from that necessity (see p. 103); likewise the bundle of shapes concept (see p. 55). A fifth unexpected advantage turned out when writing the LaTex file of this dissertation. I assumed that I had to transfer lengthy tables of repayment games from paper to LaTeX by hand. But it turned out that Mathematica can produce LaTeX code, so writing down the results became a copy-and-paste task. And finally, beyond myself, the code can be used by other researchers, too, and also save their time. So even if my initial investment was kind of high, it was worthwhile.

90 Details can be found on https://www.coursetalk.com/providers/edx/courses/spd1x-systematic-program-design-part-1-the-core-method, retrieved 2018-07-26, 15:18.

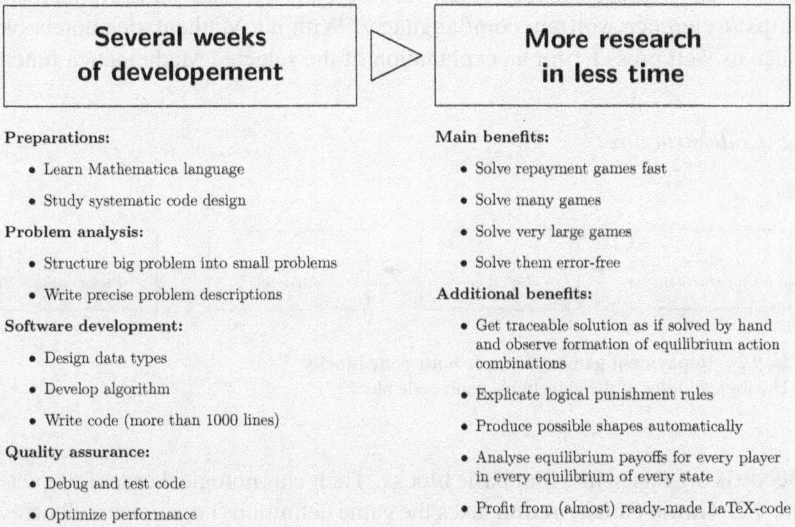

9.1 Code development challenges and benefits
A summary of the efforts to produce the software code and the resulting benefits.

9.2 Basic user manual

9.2.1 General information

The idea of this short manual is to give the uninformed reader the chance to understand the formal definitions of repayment games in the following chapters. For explanations on how exactly the algorithm works and for the documentation of the source code the reader is referred to chapter A in the appendix.

This manual does in no way replace an introduction to Mathematica, nor can it provide a documentation of the comprehensive Mathematica Language. A proficient user will have to learn the basic Mathematica principles on his own. Fortunately, the

Mathematica language is documented very well and the material is available online on https://reference.wolfram.com/language/. Within a Mathematica notebook file you can as well press F1 for an explanation of the selected Mathematica function.

9.2.2 *Code structure*

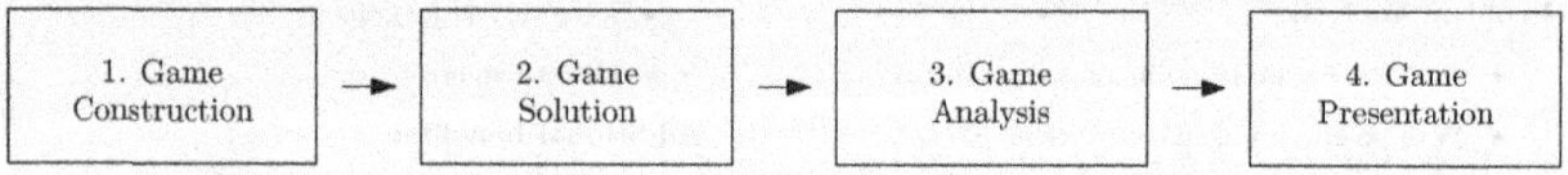

9.2 Repayment game software: Four code blocks
The logical order of the algorithms main code blocks.

The code is organised into four basic blocks. Their chronological order is depicted in figure 9.2. **Game Construction** takes the game definition (i.e. number of borrowers, action sets, payoff functions) to construct the payoff-matrix for a given shape. In addition it builds a shape specific feasibility-matrix, that tells which action combinations are in fact possible. **Game Solution** is an algorithm that can determine Nash equilibria. **Game Analysis** contains a set of functions that are needed to analyse the results of the game. It for example calculates the repayment and the loss that occurs in each shape. **Game Presentation** is a collection of functions that are busy with all kinds of graphical presentation to make the results human readable.

All these methods must be loaded into the Mathematica kernel beforehand. To get the code loaded into the Mathematica kernel open the file and execute the source code (usually by pressing `Shift+Enter`). With all these functions loaded you can then open the main file "Game Definition" where you can define the repayment game and use these methods to get the solution.

9.2.3 Important syntax

Important variable-names

`n` is the number of borrowers.

`action` is an ordered list of length three, containing an `actionName` (a string), an `actionRepayment` (a symbol) and an `actionMessage` (a string). Example: `{"2R-n",-2R,"n"}`. In games without messages `actionMessage` can be set to `""` (empty string).

`actionSet` is a list with all possible `actions` of a borrower in the unrestricted shape. Example: `{{"0R",0,""},{"1R",-1R,""},{"2R",-2R,""}}`.

`actionCombination` is a list of length `n`, with one action for each borrower. Example: `{{"2R-n",-2R,"n"},{"1R-n",-1R,"n"},{"0R-y",0,"y"}}`.

`repaymentCombination` is a list of length `n`, with the `actionRepayment` of each borrower. Example: `{-2R,-1R,0}`.

`messageCombination` is a list of length `n`, with the `actionMessage` of each borrower. Example: `{"n","n","y"}`.

Auxiliary functions related to actions

`fnGetActionName[action]`
returns the first entry of an `action`, i.e. the `actionName`.

`fnGetActionRepayment[action]`
returns second entry of an `action`, the `actionRepayment`.

`fnGetActionMessage[action]`
returns third entry of an `action`, the `actionMessage`.

`fnGetActionRepayment /@ actionCombination`
returns the `repaymentCombination` of `actionCombination`.[91]

`fnGetActionMessage /@ actionCombination`
returns the `messageCombination` of `actionCombination`.

`actionCombination[[playerKey]]` extracts from the list `actionCombination` the `action` of player number `playerKey`.[92]

91 In general: *f* /@ *expr* applies *f* to each element on the first level in *expr*. Equivalent to `Map[`*f*`,`*expr*`]`.
92 In general: *list* `[[`*n*`]]` returns the nth element of *list*. Equivalent to `Part[`*list*`,`*n*`]`.

Important Mathematica functions

`(* text *)` a comment that can be inserted anywhere in Mathematica code

Constructions

`Piecewise`[{{val_1,$cond_1$},{val_2,$cond_2$},...}] represents a piecewise defined function with values val_i in the regions defined by the conditions $cond_i$. The $cond_i$ are evaluated in turn, until one of them is found to yield `True`. The corresponding val_i is returned as value of the piecewise function. Elements of the form {val_i,`False`} are dropped, as are all elements after the first {val_i,`True`}.

`ConstandArray`[c, n] generates a list of n copies of the element c.

`$Assumptions` is the default setting for the `Assumptions` option used in such functions as `Simplify` and `Refine`.

Manipulations

`Plus @@ list` returns the sum of the entries of `list`.[93]

`Delete`[*expr*,n] deletes the element at position n in *expr*.

`Union`[*list*] gives a sorted version of *list*, with all duplicated elements dropped.

Comparisons

lhs == *rhs* returns `True` if *lhs* and *rhs* are identical.

$x <= y$ yields `True` if x is determined to be less than or equal to y.

Logic

e_1&&e_2&&... is the logical AND function. It evaluates its arguments in order, giving `False` immediately if any of the e_i is `False`, and `True` if they are all `True`. It is equivalent to using `And`[e_1, e_2, ...].

e_1||e_2||... is the logical OR function. It evaluates its arguments in order, giving `True` immediately if any of them are `True`, and `False` if they are all `False`.

!*expr* is the logical NOT function. It gives `False` if *expr* is `True`, and `True` if it is `False`.

93 In general: f @@ *expr* replaces the head of *expr* by f. Equivalent to `Apply`[f, *expr*].

9.2.4 Troubleshooting

- If calculations take too long you can press `Alt+.` to abort the evaluation of the cell.
- If evaluation fails, check whether the input makes sense. Are the action sets well-defined? Are additional assumptions needed to solve the game? Are the assumptions ambiguous or even contradictory?
- Do not delete the `rewardFunction` or the `sanctionFunction`, even if they are just defined to be zero. The algorithm will get confused if it does not find a definition (no exception handling implemented).
- Sometimes it helps to remove all variables with `ClearAll["Global‘*"]` and reload the whole code into the Mathematica kernel again. You can also shut down and restart the Mathematica kernel.

10 Perverse effect of joint liability vanishes with adaptive punishment

Abstract The perverse effect of joint liability (JL) refers to a situation where borrowers under individual liability (IL) would repay but do not do so under JL. To counteract that negative effect of JL the literature suggests more flexible lending schemes like implicit joint liability ([DQFG16]) or optimised partial joint liability ([All16]). I claim that the perverse effect is an artefact of a too rigid punishment function and I show how it disappears with adaptive punishment.

10.1 The perverse effect of joint liability

In 1995, Besley and Coate pointed at a negative effect of joint liability: "The negative effect arises when the whole group defaults, even when some members would have repaid under individual lending." [BC95, p. 1] 20 years later, de Quidt, Fetzer and Ghatak refer to this phenomenon as "perverse effect" [DQFG16, p. 219] and thereby coin the term.

We can distinguish two subtypes. The first is a variation where the lender's punishment intensity suffices to enforce IL but not JL. The group *strategically* defaults although the members would have repaid under IL.[94] In the second subtype the ability to repay suffices to repay one loan but is insufficient to bail out another borrower. The group *accidentally* defaults although some members could have repaid under IL. So either the punishment intensity or the ability to repay is insufficient for *group* repayment, nevertheless, in both cases punishment intensity and ability to repay would suffice for the repayment of *one* loan.

94 The model in [BC95] with its random punishment intensity falls into this category.

C. Prem, *The Theory of Credit Contracts*,
https://doi.org/10.1007/978-3-658-29362-8_10

10.2 Countermeasures proposed in the literature

The intention of joint liability, of course, was to increase repayment rates. Therefore researchers came up with various ideas to mitigate its perverse effect.

Besley and Coate assume that the lender has limited punishment power (see [BC95, p. 4]) and thereby encounters difficulties to enforce a joint liability contract in certain situations. But they allow for the possibility of social sanctions to target the case where a borrower refuses to contribute despite his ability to do so. Their exogenously imposed social sanction function aims exactly at these situations and is zero in all other cases (see [BC95, p. 10]). A comparison of their proposition 1 and 2 shows that with social sanctions repayment happens more often.

De Quidt et al. propose so-called implicit joint liability (IJ) as opposed to explicit joint liability (EJ). In their paper the perverse effect of joint liability is mitigated by allowing for a fallback to individual liability if necessary: "Intuitively, in those states of the world where a borrower is able to repay her loan but not that of her partner, IJ allows her to do this, yet in states of the world where she could repay her partner's loan, social capital is leveraged to encourage her to do so." [DQFG16, p. 217]

The use of social sanctions to enforce joint liability – or social capital, as it is called in [DQFG16] – is similar to [BC95]. The flexibility between individual liability and joint liability, however, is the distinctive feature. In [BC95], the type of the contract is chosen and fixed ex ante.

Other papers try to determine the degree of joint liability such that its positive effect on repayment and its perverse effect are optimally balanced. [All16], for example, interprets the perverse effect as a consequence of the burden of joint liability being to heavy. Therefore, he tries to balance out the positive risk-sharing effect of JL and the accompanying incentive for strategic default by designing contracts with an optimally determined *partial* joint liability clause.

Likewise, the so-called "flexible group lending (FGL)" in [BO10] is a contract where the amount that a borrower owes for his defaulting partner is not necessarily restricted to be the full amount of debt. Instead, the degree of joint liability is optimally determined by the bank (see [BO10, p. 350]).

10.3 Adaptive penalty functions

In this article we will see that the perverse effect of joint liability is an artefact of a rather inflexible punishment function. If full punishment sets in as soon as the group does not repay the entire amount, then there is of course no incentive to repay at least parts of the loan. The all-or-none characteristic of such punishment functions is the problem that creates the perverse effect of joint liability. Therefore I suggest the use of adaptive punishment functions. That means the punishment must be made contingent on the amount repaid. The more a borrower repays, the less ought to be punished.

To formally demonstrate how adaptive punishment works, the first challenge is to build an appropriate model. It must be complex enough to reproduce the perverse effect of joint liability, but it should do so in the simplest fashion possible, without any unnecessary assumptions. We want a model that captures the perverse effect of joint liability in its essence, without any distracting clutter.[95] Figure 10.1 demonstrates this simple setting. We have different punishment strategies and observe the resulting maps of repayment.

Individual liability provides the benchmark the definition of the perverse effect of joint liability refers to. Note that – to improve comparability to joint liability – both players have the possibility to repay $2R$ in their action sets. Under individual liability, however, the punishment strategy sets no incentive to do so.

Under joint liability with fixed punishment intensities repayment falls back behind individual liability in several equilibria – the perverse effect of joint liability. In some equilibria due to an insufficient punishment intensity to enforce group repayment, in some equilibria because of an insufficient ability to repay. We can also observe some shapes with multiple equilibria that arise due to free-riding, but they are unimportant for our analysis.

In the last repayment game the punishment strategy adapts to the amount repaid. If just $1R$ is repaid both borrowers are hit by punishment M of milder severity. As a result, the repayment in each shape is equal or even superior to individual liability – the perverse effect of joint liability disappeared! Whilst the free-riding equilibria continue to exist, we also got rid of all ambiguous cases.

---

95 For instance, assumptions about the concrete ability to repay of the borrowers, or about market structures are unnecessary as the perverse effect of joint liability is a payback-enforcement module topic (compare p. 16).

- **Individual liability** with fixed punishment intensity ($R < M$)

Punishment strategy

	P2:0R	P2:1R	P2:2R
P1:0R	-M | -M	-M | 0	0 | 0
P1:1R	0 | -M	0 | 0	0 | 0
P1:2R	0 | 0	0 | 0	0 | 0

Map of repayment

	A2:0	A2:-R	A2:-2 R
A1:0	0	R	R
A1:-R	R	2R	2R
A1:-2 R	R	2R	2R

- **Joint liability** with fixed medium punishment intensity ($R < M < 2R$)

Punishment strategy

	P2:0R	P2:1R	P2:2R
P1:0R	-M | -M	-M | -M	0 | 0
P1:1R	-M | -M	0 | 0	0 | 0
P1:2R	0 | 0	0 | 0	0 | 0

Map of repayment

	A2:0	A2:-R	A2:-2 R
A1:0	0	0 ◡	0 ◡
A1:-R	0 ◡	0 ◡ 2R	0 ◡ 2R
A1:-2 R	0 ◡	0 ◡ 2R	0 ◡ 2R

- **Joint liability** with fixed high punishment intensity ($2R < P$)

Punishment strategy

	P2:0R	P2:1R	P2:2R
P1:0R	-P | -P	-P | -P	0 | 0
P1:1R	-P | -P	0 | 0	0 | 0
P1:2R	0 | 0	0 | 0	0 | 0

Map of repayment

	A2:0	A2:-R	A2:-2 R
A1:0	0	0 ◡	2R
A1:-R	0 ◡	0 ◡ 2R	2R 2R
A1:-2 R	2R	2R 2R	2R 2R 2R

- **Joint liability** with adaptive punishment intensity ($R < M < 2R$ and $R + M < P$)

Punishment strategy

	P2:0R	P2:1R	P2:2R
P1:0R	-P | -P	-M | -M	0 | 0
P1:1R	-M | -M	0 | 0	0 | 0
P1:2R	0 | 0	0 | 0	0 | 0

Map of repayment

	A2:0	A2:-R	A2:-2 R
A1:0	0	R	2R
A1:-R	R	2R	2R 2R
A1:-2 R	2R	2R 2R	2R 2R 2R

LEGEND:
◡ Perverse effect of joint liability (group defaults although members would have repaid under IL)

10.1 The perverse effect of joint liability and adaptive punishment
Comparison of different punishment strategies and the associated maps of repayment. The perverse effect of joint liability persists unless the lender grants a reduction in punishment that outweighs the additional repayment, i.e. with $P - M > R$. The comprehensive normal forms of these repayment games are explicated and solved in appendix B.1, B.5, B.2, B.3.

10.4 Adaptive punishment in continuous games with larger groups

The basic idea behind adaptive punishment is to reward the borrower with a reduction in punishment for at least repaying parts of the loan. Let's consider a more realistic continuous case that also incorporates more than two borrowers.
How could an adaptive punishment function

$$p\colon \mathbb{R}^n \to \mathbb{R}^n,\ (R_1,\dots,R_n) \mapsto (p_i(R_1,\dots,R_n))_{i=1,\dots,n}$$

look like in this situation?

Its components, i.e. the punishment for a certain borrower i, certainly must satisfy

$$\frac{\partial p_i}{\partial R_i} \leq -1$$

to encourage the borrower to repay as much as possible. The reduction in punishment must at least outweigh the increase in repayment.

Furthermore, if $\sum_{i=1}^{n} R_i \geq nR$ we should have

$$p_i(R_1,\dots,R_n) = 0\,\forall i = 1,\dots,n$$

that is no punishment if the loan is fully repaid.

A function that satisfies these properties is

$$p_i(R_1,\dots,R_n) := \max\{nR - \sum_{i=1}^{n} R_i, 0\}\,.$$

This punishment function will of course also produce the omnipresent free-riding equilibria: if one borrower repays for the whole group the other members also escape punishment and therefore have no incentive to contribute.

10.5 Comparison to punishment functions in the literature

Let's compare the adaptive punishment strategy presented here to the punishment functions used by the other authors use in their models. In the explicit joint liability of de Quidt et al. "both contracts are renewed if and only if both loans are repaid" [DQFG16, p. 219] which is clearly a non-adaptive punishment function.

In the paper of Besley and Coate, the punishment intensity $p(\theta)$ varies with project yields θ but the group only escapes punishment if it fully repays (see [BC95, p. 3, 6]). The case that only one loan is repaid is not even considered in their game tree (see [BC95, Fig. 1])! The repayment function is again a rigid one that only distinguishes between full repayment and default.[96] However, the implicit joint liability case with its additional social sanctions in [DQFG16] and the social sanctions case in [BC95] somehow mimic a flexible punishment regimen. But social sanctions require additional incentive and information capabilities of borrowers.

Allen presents a rather complex multi-stage repayment game with a punishment strategy composed of two components (see [All16, p. 203]). One is the loss of βV and another 'joint liability'-component punishes the borrower with severity P. Certainly, these 'tools' equip the lender with some degree of flexibility concerning punishment intensity. The punishment function then goes as follows: 0 punishment for full repayment, loss of P for repayment but with a defaulting peer, and loss of βV in case of no repayment. Hence, the punishment function adapts to the amount that is repaid. Unfortunately, Allen combines this discrete punishment strategy with a continuous action set concerning the transfers between group members. Although borrowers could repay interim values according to their actions sets, what we get is borrowers that either send the amount that is necessary to bail their partner out or send nothing at all. Other transfers will not happen. The underlying reason: For a continuous action set, the three-steps discrete punishment function is not flexible enough. It simply cannot fine-tune to the transfer payment decision with enough precision. Consequently, the map of equilibrium repayments in [All16, Fig. 2.] has continuous axes but does not feature continuous repayment decisions; repayment jumps from 0 to $1R$ to $2R$. And there are still corners where the perverse effect of joint liability rules – even in the case where P is set optimally.[97]

Finally, there is a paper that uses truly adaptive punishment with a sufficiently fine-grained punishment function. In their section about ex ante side contracts Rai and Sjöström define the punishment function to be $Q_i(b_1, b_2) = [2\bar{R} - (b_1 + b_2)]/2$ (see [RS04, p. 222]). It looks similar to the punishment function defined at the end of the previous section, as it uses the difference between joint liability and joint repayment $(b_1 + b_2)$ to determine the punishment intensity. But, whilst our punishment function punishes each borrower with this difference, the punishment of Rai and Sjöström divides the difference by the number of borrowers such that

96 In the discussion section of their paper they provide a more flexible repayment game with "intermediate solutions in which borrowers agree ex post to share the burden of loan repayment in a way that is not proportional to their contractual shares." [BC95, p. 14] However, this game also only distinguishes between $r_1 + r_2 \geq 2r$ and $r_1 + r_2 < 2r$.

97 In region V in [All16, Fig. 2.] JL repayment is below IL repayment.

each borrower only carries a fraction of it. This is in conflict with our finding that the derivative must be less or equal -1, but it works here as Rai and Sjöström assume fully enforceable side contracts and hence the agents maximize their *joint* surplus. Joint punishment is $2\bar{R} - (b_1 + b_2)$ and everything is fine, no perverse effect appears.[98] However, to rely on fully enforceable side contracts is a very strong assumption. Without perfect side contracting, this punishment function would not work any more, because if a borrower repays an additional amount, the reduction of his punishment would only be an nth of this amount. For the case where borrowers cannot side contract, Rai and Sjöström present a cross-messaging mechanism instead that avoids the perverse effect of joint liability ([RS04, Proposition 3]). But still, if borrowers do not have the information about each other's ability to repay, then this mechanism also ceases to exist.[99]

Compared to all these attempts to tackle the perverse effect of joint liability, our adaptive punishment function, however, is modest and robust and attends to its duty to implement joint liability whilst avoiding the perverse effect. Its simplicity and its frugality concerning prerequisites make adaptive punishment very expedient:

- We do not rely on additional information (like [DQFG16] do, where the borrowers use their information to implement implicit joint liability by mutually ensuring each other). It is not necessary to have information about the state of a borrower's ability to repay to apply the adaptive punishment function.
- We do not need increased punishment capabilities compared to ordinary joint liability (which was the point of social sanctions in [BC95]). Let $M = R + \varepsilon$ and $P = R + M + \varepsilon$ for an infinitesimally small $\varepsilon > 0$, then all conditions are satisfied and $P = 2R + 2\varepsilon$ is just infinitesimally larger than $2R$ – as it was in the third repayment game of figure 10.1. (Almost) no additional punishment capabilities are necessary.
- We do not need to weaken the joint liability clause (as with partial joint liability in [All16]). The adaptive punishment function implements full joint liability.

Without the need for more information, more punishment or a weakened JL clause, we still get a higher repayment compared to IL and compared to JL. In addition, the adaptive punishment contract has a lower overall punishment load than the original JL contract (compare appendix B.2 and B.3), so it is even superior from a welfare perspective.

98 It is basically an application of [Dia84], see p. 56.

99 Cross-reporting mechanisms and side-contracting are very efficient with respect to equilibrium punishment load, see chapter 12. If feasible, apply them instead of using adaptive punishment.

10.6 Summary and discussion

In a nutshell: The perverse effect of joint liability is a 'threshold-effect' of stepped punishment functions. If the borrowers can control their repayment more precisely than the punishment function can react, the perverse effect of joint liability will be an issue.

That leaves the question whether this is a theoretic problem or a real problem. Did researchers neglect to model punishment functions fine-grained enough? Or are our incentive tools (see p. 27) in fact that baulky?

Our first intuition says that the often used (partial) exclusion from financial markets can be fine-tuned quite easily by gradually worsening conditions for future loans (e.g. higher interest, smaller loan size) or delaying disbursement. However, imagine a borrower that needs money at a certain point in time.[100] A little delay is not a punishment as long as the loan is disbursed early enough. But delaying disbursement to a point after the opportunity has passed, suddenly depletes some of the borrowers potential utility. Although in theory delaying is a gradual process, its effective impact may be subject to quantization. On the other hand, making loans more expensive via increasing interest rates seems to be a completely gradual process.

If we look at empirical results, repayment rates do not drop when switching from joint liability to individual liability (see [GK14]). The interpretation could be that the insurance effect and the perverse effect of joint liability balance each other out. But what about joint liability with adaptive punishment? Here – according to our analysis – a change to individual liability should in fact decrease repayment rates. Is adaptive punishment not being used in reality?

The question, whether the perverse effect is a real issue or just appears in theoretic models that are built too coarsely is awaiting an answer. At this juncture, the only advice that can be given to a lender is to act with caution if your incentive tools cannot be adequately fine-tuned. A hurdle that, in some situations, appears to be unattainably high, will entirely lose its intended incentive effect in these situations.

100 This could be a once in a lifetime investment opportunity, paying for a life-saving medical treatment, or any other opportunity with an expiry date.

11 Social sanctions coordinate and motivate

Abstract This article examines the effects of social sanctions on various repayment games. Compared to the model in [BC95] the avoidable-harm-concept of social sanctions is kept, but the setup is simpler and cleaner. And yet, the two main roles of social sanctions can be clearly demonstrated: its incentive effect and its coordination effect. Moreover, the model presented here reaches beyond the results in [BC95] in several ways. Firstly, all shapes are considered ([BC95] only consider the unrestricted shape). Secondly, group size is increased to three ([BC95] only analyse groups with two borrowers). Thirdly, social sanctions are combined with individual liability ([BC95] only examine social sanctions in conjunction with joint liability).

11.1 The difficulties in the model of Besley and Coate

Social sanctions have been modelled by Besley and Coate in their publication "Group lending, repayment incentives and social collateral" from 1995 ([BC95]). They did pioneer work in the analysis of social sanctions and their famous paper has already been cited 274 times.[101] Pioneering is always difficult and chances that future research will suggest improvements are high. This section will talk the reader through some features of the model of Besley and Coate and describe how the model presented here differs.

Even though they claim to use "the simplest possible model to make the main point of interest" ([BC95, p. 3]) they use a comparatively complex two stage sequential repayment game to model joint liability (see [BC95, Fig.1&2]). The contributing borrower is given the chance to react to the outcome of a simultaneously played contribution-game, whilst the non-contributing borrower does not get the chance to revise. In the model presented here we use a simple one-stage repayment game instead.

In the model of [BC95] some repayment combinations remain unconsidered (see figure 11.1). Whilst ruling out the combinations where joint repayment exceeds $2R$ seems reasonable, ignoring the possibility that only one loan is repaid feels awkward.

101 According to `http://citec.repec.org/cgi-bin/get_data.pl?h=RePEc:eee:deveco:v:46:y:1995:i:1:p:1-18&o=all`, 2016-10-14, 13:30

C. Prem, *The Theory of Credit Contracts*,
https://doi.org/10.1007/978-3-658-29362-8_11

	P2:0R	P2:-1R	P2:-2R
P1:0R	✓	×	✓
P1:-1R	×	✓	×
P1:-2R	✓	×	×

11.1 Repayment combinations considered in [BC95]
In [BC95] the case that only one loan gets repaid is completely ignored. The borrowers' action sets exclude such decisions.

It would be ok to show that these combinations are never played. But it is flawed logic to exclude them from the set of possible strategy combinations right away. In contrast, the repayment game presented here considers all repayment combinations.

Besley and Coate also neglect some shapes of their repayment game and say they "ignore the possibility that a borrower has insufficient funds to repay his loan." [BC95, p. 4, footnote 7] In fact, they focus on a single shape only and solely ask if a borrower does repay, but not if he can repay. Accidental default is not even considered. Consequently, the performance of their scheme in these situations remains in the dark. Of course, we will solve all shapes.

	A2:0R	A2:1R	A2:2R
A1:0R	×	×	×
A1:1R	×	×	×
A1:2R	×	×	✓

11.2 Ability combinations considered in [BC95]
In [BC95] they focus on the unrestricted shape and ignore the performance of their repayment game in other situations.

In [BC95] θ is the income of an investment project. But, as we have just seen, the project yield does not restrict the ability to repay and shape the repayment game, instead it determines the penalty $p(\theta)$ in their model – which is a rather cumbersome way to say available punishment intensity is a random variable. This random variable is then modelled continuously. At the same time borrowers are left with only discrete action sets – they can either repay 0 or r or $2r$ [BC95, p. 6]. It turns out that the only relevant information about the distribution are the two probabilities $F(\phi(r))$ and $F(\phi(2r))$ and their respective complementary probabilities (see [BC95, Eq.(1),(2),(3)]), which are defined by the thresholds $\phi(r)$ and $\phi(2r)$ (see [BC95, p. 4]).

The actually interesting feature of the punishment function in Besley's and Coate's model is that there are situations where the incentive is too weak to enforce repayment [BC95, p. 4]. In these cases, social sanctions can help to push the overall incentive to repay over a critical threshold. We will model incomplete bank sanctions by simply calculating the outcome of the repayment game twice: Once in case of a (non-stochastic) high punishment intensity and once for a (non-stochastic) weak punishment intensity. Thereby we can examine and observe the effects of social sanctions in the two relevant cases without the hassle of a stochastic punishment function.

Finally, whilst Besley and Coate tweak their results by assuming away group default in ambiguous cases, we will simply state cases with multiple equilibria.[102] This procedure displays worst cases and best cases in absolute transparency.

11.2 The two components of social sanctions

Although the goal is to build a cleaner and clearer model of social sanctions than that in [BC95], the core principles and concepts shall be kept. This includes

- the fundamental properties of the social sanction function and
- the fact that the lender's punishment intensity may be restricted.

The second item will be considered simply by calculating the repayment game once with high punishment intensity $2R < P$ and once again with medium intensity $R < M < 2R$.

Emulating the first item requires requires some more thinking. Besley and Coate describe social sanctions as an (exogenously given) social rule with two key properties: "First, penalties depend upon the extent of harm inflicted by the non-contributing member on his partner and second, they depend upon the reasonableness of the decision not to contribute." [BC95, p. 10] Accordingly, their assumption 1 captures the following principles:

i) No harm $\Rightarrow$ no sanctions

ii) No ability to help $\Rightarrow$ no sanctions

iii) The more avoidable harm $\Rightarrow$ the more sanctions

102 See p. 65 on why this transparency is important.

So sanctions are applied in cases of unnecessary harm. Let's dig deeper into the analysis of avoidable harm. When does it occur? The possibilities where a borrower suffers avoidable harm – and consequently social sanctions are applied – are:

Not-contributing. If the group does not fully repay, the borrowers will *suffer* from punishment for default. If full repayment was in fact possible and a borrower $\mathcal{B}$ contributed less than he actually could, the punishment is *avoidable*. So if $\mathcal{B}$ refuses to help he should be sanctioned with S_n for not contributing enough.

Free-riding. If borrower $\mathcal{A}$ repays for his partner he *suffers* from additional repayment. If his partner $\mathcal{B}$ could in fact repay on his own, this is *avoidable*. So if $\mathcal{B}$ refuses to repay his own loan he should be sanctioned with S_f for free-riding on his peer.

These cases are also reflected by the piecewise defined loss function $L(\theta, r) \equiv \min\{p(\theta) - r, r\}$ in [BC95, p. 10]. The non-contributing borrower causes avoidable harm either by triggering unnecessarily high punishment or via provoking an unnecessary bail-out. Let R_i be the amount repaid by borrower $\mathcal{B}_i$ and A_i the realisation of his ability to repay. Then the rules above can be formalised the following way:

Not-contributing. $(\Sigma_{j=1}^{2} R_j < 2R) \wedge (\Sigma_{j=1}^{2} A_j \geq 2R) \wedge (R_i < A_i) \Longrightarrow -S_n$ for $\mathcal{B}_i$

Free-riding.[103] $(\Sigma_{j=1}^{2} R_j \geq 2R) \wedge \quad (A_i > 0) \quad \wedge \; (R_i = 0) \Longrightarrow -S_f$ for $\mathcal{B}_i$

Q 11.1 Mathematica code of the social sanctions rule under JL

The code below shows how the social sanction rules are implemented in Mathematica.

```
(* SANCTIONS FUNCTION: borrower to deadweight flow ---- *)
sanctionFunction [ abilityCombination_ ,actionCombination_,playerKey_]:=Refine[Piecewise[{
(* not-helping condition *)
{-Sn,   -Plus@@(fnGetActionRepayment/@actionCombination)<n R (* not all loans are repaid *)
&& -Plus@@abilityCombination>=n R (* full repayment is possible *)
&& -fnGetActionRepayment[actionCombination[[playerKey]]]<-abilityCombination[[playerKey]](* player has resources *)},
(* free-riding condition *)
{-Sf,   -Plus@@(fnGetActionRepayment/@actionCombination)>= n R (* all loans are repaid *)
&& -abilityCombination[[playerKey]]>=R (* can repay his loan *)
&& fnGetActionRepayment[actionCombination[[playerKey]]]==0 (* does not repay his loan *)},
{0,True} (* default condition *) }]];
```

Figure 11.3 illustrates what the sanction component of the borrowers' payoff function looks like if these rules are applied.

103 Note that, logically, free-riding borrowers should also be punished if $\Sigma_{j=1}^{2} R_j > 2R$, although this case will never occur.

		A2:0			A2:-R			A2:-2R		
		P2:0R	**P2:1R**	**P2:2R**	**P2:0R**	**P2:1R**	**P2:2R**	**P2:0R**	**P2:1R**	**P2:2R**
A1: 0	**P1:0R**	0 | 0	×	×	0 | 0	0 | 0	×	0 | -Sn	0 | -Sn	0 | 0
	P1:1R	×	×	×	×	×	×	×	×	×
	P1:2R	×	×	×	×	×	×	×	×	×
A1: -R	**P1:0R**	0 | 0	×	×	-Sn | -Sn	-Sn | 0	×	-Sn | -Sn	-Sn | -Sn	-Sf | 0
	P1:1R	0 | 0	×	×	0 | -Sn	0 | 0	×	0 | -Sn	0 | 0	0 | 0
	P1:2R	×	×	×	×	×	×	×	×	×
A1: -2R	**P1:0R**	-Sn | 0	×	×	-Sn | -Sn	-Sn | 0	×	-Sn | -Sn	-Sn | -Sn	-Sf | 0
	P1:1R	-Sn | 0	×	×	-Sn | -Sn	0 | 0	×	-Sn | -Sn	0 | 0	0 | 0
	P1:2R	0 | 0	×	×	0 | -Sf	0 | 0	×	0 | -Sf	0 | 0	0 | 0

11.3 Social sanctions payoff component in case of joint liability
This table shows the social sanctions component of the borrowers' payoff under joint liability with high punishment intensity. It is defined by the sanction rules on p. 170. S_n is the sanction for not contributing and S_f that for free-riding.

Now that we know *when* social sanctions are applied, let's talk about the intensity of social sanctions. Besley and Coate use the loss function mentioned above to determine the sanction intensity. Their assumption (iii) says that the intensity should increase with the amount of inflicted harm as well as with the degree of avoidability. However, social sanctions will only be effective if their intensity surpasses a certain threshold. The reason is that we – as Besley and Coate do – use discrete borrower action sets. Since we want to examine the effect of social sanctions, we assume that they are effective, i.e. $S_n > R$ to motivate the borrower to bail out his partner and pay an additional R, and $S_f > R$ to keep borrowers from free-riding.

11.3 The effect of social sanctions on joint liability lending

To observe the effect of social sanctions, in this section we will compare repayment games with and without them.

11.3.1 Coordination effect

Let's first of all take an explicit joint liability contract with high punishment intensity $2R < P$. If we look at the map of equilibria without social sanctions we encounter some free-riding equilibria. Since free-riding is discouraged by social sanctions they disappear in the presence of social sanctions.

JL – without social sanctions

	A2:0	**A2:-R**	**A2:-2R**
A1: 0	0R | 0R	0R | 0R	0R | 2R
A1: -R	0R | 0R	0R | 0R 1R | 1R	0R | 2R 1R | 1R
A1: -2R	2R | 0R	1R | 1R 2R | 0R	0R | 2R 1R | 1R 2R | 0R

JL – with social sanctions

	A2:0	**A2:-R**	**A2:-2R**
A1: 0	0R | 0R	0R | 0R	0R | 2R
A1: -R	0R | 0R	1R | 1R	1R | 1R
A1: -2R	2R | 0R	1R | 1R	1R | 1R

11.4 Coordination effect of social sanctions
The map of equilibria for a joint liability loan high punishment intensity ($2R < P$), without and with social sanctions. For details, see B.2 and B.4.

Multiple equilibria are problematic (see p. 65). Coordination failures may lead to situations where not even an equilibrium is played. Social sanctions – in particular the anti-free-riding-component – help the borrowers to agree on a particular equilibrium. That's the *coordination effect* of social sanctions. For the lender, however, the maps of repayment show no change (compare B.2 and B.4). That is the case since the lender's punishment capabilities suffice to enforce the repayment of $2R$ anyway.

By the way, with adaptive punishment the not-contributing sanction rule must be changed to

$$\left(\sum_{j=1}^{2} R_j < 2R\right) \bigwedge (R_i < A_i) \Longrightarrow -S_n \text{ for borrower } i$$

since also any partial repayment reduces (avoidable) lender punishment. The effect is the same, namely that the free-riding equilibria in B.3 vanish.

11.3.2 Incentive effect

More similar to the model of Besley and Coate is the case where the lender's punishment capability is incomplete, i.e. we have a medium punishment intensity $R < M < 2R$. Compared to the case of high punishment intensity the map of equilibrium repayments deteriorates drastically. However, in the presence of social sanctions the lender's punishment is leveraged and leads to a quite satisfying map of repayment.

JL – without social sanctions

	A2:0	**A2:-R**	**A2:-2R**
A1: 0	0	0	0
A1: -R	0	0 2R	0 2R
A1: -2R	0	0 2R	0 2R

JL – with social sanctions

	A2:0	**A2:-R**	**A2:-2R**
A1: 0	0	0	2R
A1: -R	0	2R	2R
A1: -2R	2R	2R	2R

11.5 Incentive effect of social sanctions
The map of repayment for a joint liability loan with medium punishment intensity ($R < M < 2R$), without and with social sanctions. For details, see B.5 and B.6.

The effect of social sanctions in this repayment game is astonishing. Not only are the ambiguous cases gone, in addition the repayment in other shapes of the game increase.[104] Borrowers bail each other out despite the low intensity of the lender punishment. That's the *incentive effect* of social sanctions. The intuition behind this result is that punishment by the lender works like a trigger that induces social sanctions by the borrowers in cases where lender punishment is avoidable. Thereby social sanctions can leverage incentives set by the creditor.

If the lender is certain that borrowers can exert social sanctions (i.e. have the information and incentive tools to do so) he may even *intentionally* reduce his punishment intensity from P to M in order to lower equilibrium punishment load and thus increase welfare (compare B.4 and B.6). Social sanctions do not add to the punishment load as they are never used in equilibrium.

11.3.3 Incentive effect with three borrowers

In this section we will try and apply the social sanction function to a joint liability game with three borrowers. Which adaptations do we have to make? The sanction rules are readily extended to several borrowers.

Not-contributing. $(\sum_{j=1}^{n} R_j < nR) \bigwedge (\sum_{j=1}^{n} A_j \geq nR) \bigwedge (R_i < A_i) \Longrightarrow -S_n$ for $\mathcal{B}_i$

Free-riding. $(\sum_{j=1}^{n} R_j \geq nR) \bigwedge \quad (A_i > 0) \quad \bigwedge \ (R_i = 0) \Longrightarrow -S_f$ for $\mathcal{B}_i$

The motivation created by the sanctions, i.e. its intensity, should be increased to keep them effective. We now assume that $S_n > (n-1)R$ to motivate a borrower to even bail out all of his peers if necessary and possible.[105] $S_f > R$ can stay that way to motivate a borrower to repay his own loan, if he can.

Let's have a look at an *excerpt* of the resulting map of equilibria (figure 11.6). Notice the shape with multiple equilibria. Fortunately, it is not an ambiguous case, since both equilibria amount to the same joint repayment. But somehow it is unclear who bails the first borrower out – a new coordination challenge. How does that come about? The reason is twofold. First of all, in this game it could be an artefact of the definition of the borrowers' action sets. The possibility to repay $1.5R$ and equally share the burden simply does not exists. When we extend the actions sets we get an additional equilibrium with actions $0R|1.5R|1.5R$, but the other two equilibria do not vanish. Therefore the coordination problem must be an issue of the social

104 A similar result with increased group repayment in the presence of social capital has been discovered by Arnold et al. in [ARS13]: Compare figure 1 (p. 68) with figure 3 (p. 80).

105 The increased sanction intensity only gets effective if we also extend the borrowers' action sets by the possibility to repay more than two loans. We do not do it here, but it is an easy change in the Mathematica code of the game and the results are alike.

sanction function we defined. And in fact, there is no sanction rule in our sanction function that increases the attractiveness of one of these equilibria.

A1:0

	A3:0	A3:-R	A3:-2R
A2:0	0R | 0R | 0R	0R | 0R | 0R	0R | 0R | 0R
A2:-R	0R | 0R | 0R	0R | 0R | 0R	0R | 1R | 2R
A2:-2R	0R | 0R | 0R	0R | 2R | 1R	0R | 1R | 2R 0R | 2R | 1R

11.6 Coordinating social sanctions in larger groups
This table shows an excerpt of the map of equilibria for a joint liability contract with medium punishment intensity ($R < M < 2R$) with social sanctions. Details in B.7.

To distribute the bail-out burden evenly amongst helping peers we need an additional sanction rule. What could such anti-shirking rule look like? It could for example measure the ratio $\frac{R_i}{A_i}$ and punish the borrowers that contribute proportionally less to bail-out activities.

Another question that is not modelled here is the question who exactly sanctions whom. In a group of two there is no doubt. But in larger groups responsibility is diffuse. Altogether, we must conclude that although it seems as if increasing group size is just a matter of more borrowers, it actually raises new problems and requires additional concepts and assumptions. Against that background it is highly questionable if all the group lending results established for a pair of two borrowers easily extend to larger groups.

11.4 The effect of social sanctions on individual liability lending

Wouldn't it be interesting to combine social sanctions with individual liability? The goal of this section is to examine the outcome of the repayment game in such a situation. As before, the first challenge consists of interpreting the avoidable-harm-paradigm of Besley and Coate in this new environment. Which are the cases where avoidable harm occurs?

		A2:0			A2:-R			A2:-2R		
		P2:0R	P2:1R	P2:2R	P2:0R	P2:1R	P2:2R	P2:0R	P2:1R	P2:2R
A1: 0	P1:0R	0 | 0	×	×	0 | 0	0 | 0	×	0 | -Sn	0 | -Sn	0 | 0
	P1:1R	×	×	×	×	×	×	×	×	×
	P1:2R	×	×	×	×	×	×	×	×	×
A1: -R	P1:0R	0 | 0	×	×	0 | 0	0 | 0	×	0 | 0	0 | 0	0 | 0
	P1:1R	0 | 0	×	×	0 | 0	0 | 0	×	0 | 0	0 | 0	0 | 0
	P1:2R	×	×	×	×	×	×	×	×	×
A1: -2R	P1:0R	-Sn | 0	×	×	0 | 0	0 | 0	×	0 | 0	0 | 0	0 | 0
	P1:1R	-Sn | 0	×	×	0 | 0	0 | 0	×	0 | 0	0 | 0	0 | 0
	P1:2R	0 | 0	×	×	0 | 0	0 | 0	×	0 | 0	0 | 0	0 | 0

11.7 Social sanctions payoff component in case of individual liability
This table shows the social sanctions component of the borrowers' payoff under of individual liability. It is defined by the sanction rules on p. 177. It tells when a borrower is punished with social sanctions S_n.

If borrower $\mathcal{B}_j$ cannot repay he will default and *suffer* from punishment for default. But if borrower $\mathcal{B}_i$ has the ability to bail him out it makes the punishment *avoidable*. If $\mathcal{B}_i$ does not bail his partner out he should be punished with $-S_n$ for not-helping.

$$(A_j = 0) \bigwedge (A_i \geq 2) \bigwedge (R_i < 2) \Longrightarrow -S_n \text{ for } \mathcal{B}_i$$

Note that in reality we could imagine a situation where borrower $\mathcal{B}_i$ decides to help his peer but default on his own loan. In our model this is not possible since the action sets do not differentiate between whose loan is repaid. We could enrich the action sets in two ways to capture that possibility. Either each borrower gets actions for repaying is own loan *and* actions for repaying the loan of his peer, or equivalently we allow actions that represent transfers between borrowers. For the rest of the section, however, a borrower – by the design of the game – has to repay his own loan first, before being allowed to bail his partner out by repaying an additional R. Without joint liability, there is no free-riding (see equilibria in B.1). If player $\mathcal{B}$ can repay his loan but does not do so, it does not inflict additional punishment on $\mathcal{A}$ since there is no joint liability. And 'no harm' in particular means there can be no 'avoidable harm'.

Q 11.2 Mathematica code of the social sanctions rule under IL

The code below shows how the social sanction rules are implemented in Mathematica.

```
(* SANCTIONS FUNCTION: borrower to deadweight flow ----- *)
sanctionFunction [ abilityCombination_ ,actionCombination_, playerKey_]:=Refine[Piecewise[{
(* not-helping condition *)
{-Sn,     MemberQ[Delete[abilityCombination,playerKey],0] (* peer cannot repay *)
&&-abilityCombination[[playerKey]]>R (* borrower has more resources than he needs for his own *)
&&-Plus@@(fnGetActionRepayment/@actionCombination)<n R (* borrower does not help *) },
{0,True}
}]];
```

The results of individual liability with social sanctions are quite impressive. If borrowers sanction each other for not helping each other, repayment rates are at their best. Only joint liability with adaptive punishment delivered similar repayment performances (compare B.8 and B.3).

This is in accordance with an empiric result by [GK14] which shows that repayment rates did not drop when switching from joint liability to individual liability. In the presence of social sanctions repayment rates may even increase (compare B.8 to B.4 and B.6).

An even closer look at the results of adaptive joint liability and individual liability with social sanctions reveals that punishment load is lower in the latter (compare B.8 with B.3). What is the intuition behind this effect? Certainly, repaying two loans requires some motivation. Under individual liability with social sanctions

IL – without social sanctions

	A2:0	A2:-R	A2:-2R
A1:0	0	R	R
A1:-R	R	2R	2R
A1:-2R	R	2R	2R

IL – with social sanctions

	A2:0	A2:-R	A2:-2R
A1:0	0	R	2R
A1:-R	R	2R	2R
A1:-2R	2R	2R	2R

11.8 Incentive effect of social sanctions for individual liability
The map of repayment for individual liability, without and with social sanctions. For details, see B.1 and B.8.

some of this motivation comes from the borrowers themselves. Borrowers, however, can target their incentives much more precisely than the lender: Social sanctions are shape dependent, whilst lender punishment must be the same in each shape due to hidden states. As a result social sanctions are precisely targeted at the avoidable harm spots but are never exerted in equilibrium – which saves dead-weight loss.

Maybe the reader has asked himself or herself: Why should borrowers actually commit to helping each other out without joint liability? This question can also be answered by looking at punishment load. Compare B.8 and B.1 to see that punishment load with social sanctions is even lower than under pure individual liability. Punishment by the lender can be avoided by helping each other in the (A1:0|A2:-2R) shape and its symmetric counterpart.

The results presented here are also in line with those of de Quidt et al. They interpret social sanctions as "the net present value of lifetime payoffs in a repeated "social game" played alongside the borrowing relationship" [DQFG16, p. 222], as opposed to Besley and Coate's concept where social sanctions depend on avoidable harm and project success. In their intricate model they compare individual liability with implicit joint liability and explicit joint liability and also arrive at the superiority of implicit joint liability if evaluated from a welfare perspective. As de Quidt et al. mention, this may rationalise the observed shift away from the use of explicit joint liability by microfinance institutions ([DQFG16, p. 217]).

Finally, note that the social sanction rule presented in figure 11.7 provides a concrete formal explication how perfect side-contracting – like the perfect side-contracting capability of borrowers assumed in [RS13] – can indeed be realised and thereby backs up the plausibility of their assumption.

11.5 Summary and further research

We analysed the effects of social sanctions on a number of different repayment games. The analysis completely took place within the payback-enforcement module of lending theory. To define the social sanction rules, the avoidable-harm-concept of [BC95] has been carried over. However, we distinguished two conceptually different instances of avoidable harm and correspondingly two different instances of social sanctions: for not helping and for free-riding.

In the games with social sanctions we observed two effects: firstly, a coordination effect towards a unique equilibrium, and secondly an incentive effect that increases repayment rates. In addition we found that repayment games with social sanctions are more efficient concerning punishment load.

Altogether, we can derive two pieces of advice. The advice for creditors is to prefer individual liability with social sanctions over joint liability with adaptive punishment – if feasible, i.e. if borrowers have the necessary information and incentive tools available. Repayment rates will be maximal with both payback-enforcement schemes but the equilibrium punishment load will be less under individual liability with social sanctions. The advice for debtors is to close mutual insurance agreements to save punishment by the bank.

Papers about social sanctions like [BC95, DQFG16] and this chapter here typically make some sophisticated assumptions about the sanction rules the borrowers apply. However, it would be interesting what are the actual rules borrowers use in real life. Is avoidable harm the criterion that is used in reality?

Another empirical question is related to the functioning of the incentive tools borrowers use for sanctioning their peers. How is the borrowers' sanctioning intensity restricted? Is lending group exclusion the maximum penalty or can they hassle each other in their every day life? The intensity of some sanctioning tools may depend on the individuals relative social rank. It may be difficult for an inferior member of a group to threaten the leader with exclusion. So the leader might be able to enforce his bail-out whilst an inferior group member may have no such power. Such settings would require theoretic research about asymmetric repayment games. It is also unclear how the social sanctions of several group members add up to the final sanction intensity. And the answers to all these questions are essential for lenders in order to evaluate if borrowers are able to close mutual insurance agreements or not. If they are not, the lender should switch to adaptive joint liability (see chapter 10).

Concerning future theoretic research on social sanctions it would be nice to extend all the repayment games mentioned in this article to group sizes of $n > 2$. In some cases this is rather tricky as the responsibilities for avoidable harm get blurred. As we have seen it will also be necessary to enrich the action sets of the borrowers with non-integer multiples of R. Maybe even continuous action sets should be considered to examine how exactly the burden of bail-out is shared amongst group members. So social sanctions models for groups of more than three borrowers needs a lot more assumptions. Without further knowledge or further research on how sanctions work in reality, these assumptions would be rather arbitrary. A joint research program together with social psychologists seems advisable to gain more insight.

12 Cross-reporting can eliminate equilibrium punishment load

Abstract This article analyses the effects of cross-reporting on repayment games. The joint liability message game in [RS13] is presented and corrected. The joint liability message game in [RS04] is stated and extended to a richer set of states of nature. Finally, a new cross-reporting game in combination with individual liability is presented that features zero equilibrium punishment load and only unique equilibria for all realisations of a three-valued ability to repay.

12.1 Combining repayment with a message

The Microfinance literature comprises two papers ([RS04] and [RS13]) that combine repayment games with message games. Both papers, by the way, are co-authored by Sjöström, an expert in implementation theory.[106] The basic assumption is that borrowers in a borrowing group have information about the realisation of each other's ability to repay. Why not simply ask them to reveal the state of their peers? Of course, the incentives must be cleverly set to motivate borrowers to be honest.

12.2 Cross-reporting and joint liability in [RS13]

Let's first have a look at the mechanism in their publication of 2013:

"Consider the following message game. The bank asks each villager whether they can repay $2(1+r^*)$ in full, and each villager responds "yes" or "no". Simultaneously, each villager i makes a repayment b_i to the bank. (1) If both said "yes", an the loan is in fact repaid in full, $b_1+b_2=2(1+r^*)$, then neither is punished. (2) If there are no repayments ($b_1=b_2=0$), then anyone who said "no" escapes punishment, but anyone who said "yes" is punished by a large amount (e.g. denied all future loans). (3) If villager i said "no" and made no repayment ($b_i=0$), but villager j said "yes" and repaid $b_j=2(1+r^*)$, then villager i is punished by a large amount, while villager j receives a reward: the repayment b_j is returned to her, plus a small "bonus" $\varepsilon>0$. (4) In all other cases: both villagers are punished by a large amount." [RS13, p. 259, footnote 11]

106 Sjöström is one of the authors of the chapter about implementation theory in the "Handbook of Social Choice and Welfare" [MS02].

C. Prem, *The Theory of Credit Contracts*,
https://doi.org/10.1007/978-3-658-29362-8_12

Notice that the action set of a borrower is richer than in an ordinary repayment game. Formally it is the set {0R-y,0R-n,1R-y,1R-n,2R-y,2R-n}.[107] In addition to a certain repayment, both borrowers send a message. However, "1R-n" and "2R-n" are always dominated by "0R-n" thus can be cancelled. Besides, a strategy that combines a repayment of 2R with the message "no" logically does not make much sense. Thus, the message in Rai and Sjöström's game is actually only prompted if both borrowers repay 0R. The underlying idea is to be able to distinguish strategic default and accidental default – and abate punishment in the later. Also note that in the game description above a villager is asked to report about their *joint* ability to repay. However, since they are only two, together with the villager's own repayment the information about his peer's state can be derived.

If we carefully stick to the description of these rules, the resulting repayment game exhibits shapes in which no equilibrium exists (see B.9). My supervisor Prof. Dr. Lutz Arnold pointed out that there is a fix, if their rule (2) is slightly improved. The necessary change is: If there are no repayments and they do not consistently send the message "no", everybody is punished – not only the borrower who said "yes". The following tables illustrate the outcomes of the resulting *improved* version.

Map of lender payoffs

	A2:0	**A2:-R**	**A2:-2R**
A1: 0	0 0	0 0	2R
A1: -R	0 0	0 0 2R	2R 2R
A1: -2R	2R	2R 2R	2R 2R 2R

Map of deadweight loss

	A2:0	**A2:-R**	**A2:-2R**
A1: 0	0 -2P	0 -2P	0
A1: -R	0 -2P	0 -2P 0	0 0
A1: -2R	0	0 0	0 0 0

12.1 Results for the corrected version of the message game in [RS13]
The tables show the map of lender payoff (i.e. repayment received minus rewards paid) and the map of deadweight loss for the corrected version of the message game in [RS13]. For details see B.10.

The nice feature of this game: If borrowers choose 'the right' equilibria, we get the repayment performance of a typical joint liability contract but without any equilibrium punishment load.

107 $R = (1 + r^*)$ is the stipulated return service of one borrower.

However, let's thoroughly analyse these multiple equilibria. For the west, northwest and north shape the equilibria conform in repayment but differ in the amount of punishment. So a coordination mechanism is necessary to reach the equilibrium with zero punishment. Given borrowers have access to such coordination means, they should easily agree on playing the better equilibrium. Having said that, consider the centre shape. Beside the desired Nash equilibrium, it includes two additional ones with zero repayment. If borrowers are able to coordinate, they can coordinate towards the most attractive {0R-n,0R-n} action combination. So this shape is vulnerable to collusion.[108]
Altogether we must conclude: *Without* cooperation, coordination failure may impede the intended reduction of equilibrium punishment load. But *with* cooperation, we get a drop in repayment rates due to collusion. The nice feature of this game does not really emerge.

12.3 Cross-reporting and joint liability in [RS04]

12.3.1 Original version

Let's turn to the 2004 paper of Rai and Sjöström and inspect the cross-reporting mechanism presented there. First of all we translate their notation into ours.

In [RS04]	Meaning (see page in [RS04])	Here
$y_i \in \{0,h\}$	borrower *i*'s investment project output (p. 220)	none
$\Psi > h$	maximum feasible punishment (p. 220)	P
$\bar{R}$	face value of debt (p. 222)	R
$0 < \varepsilon < \Psi - h$	potential reward (p. 226)	e
$(0,\mu)$	action: repay nothing and report peer success (p. 226)	"0R-y"

12.2 Notation used in [RS04]
For the convenience of the reader, the table explains what the symbols used in [RS04] mean and what its respective counterparts in our notation are.

Note that their two-valued ability to repay can *only* be h or 0 which accounts to only four states of the world: $(0,0),(0,h),(h,0),(h,h)$. Situations where both borrowers have a positive ability to repay but of different magnitude are ignored. Consequently, cases where a borrower can chose from an action set of the form $\{0,(0,\mu),\bar{R}\}$ are inexistent in their proof of proposition 3 (see [RS04, p. 227 f.]).

108 [RS13] themselves remark on page 295 in footnote 11 that the message game feature of their repayment game renders the game vulnerable to collusion between borrowers.

	A2: 0	A2: $-\bar{R}$	A2: $-h$
A1: 0	✓	×	✓
A1: $-\bar{R}$	×	×	×
A1: $-h$	✓	×	✓

12.3 Ability combinations considered in [RS04]
The ability combinations (i.e. shapes) considered in [RS04]. Situations where both borrowers have a positive ability to repay but of different magnitude are ignored. Also the middle shape is missing as by their assumptions h is always greater than $\bar{R}$.

We will increase the model's realism by extending it to a three-valued ability to repay. As usual, we focus on the three relevant realisations for discrete joint liability repayment games, namely borrowers that can either repay nothing **0R**, or just their own loan **1R**, or both loans **2R.**[109] Next, we reproduce the punishment function and reward function defined in [RS04, p. 226]. This time, instead of giving verbal rules (compare p. 181), they explicitly state the reaction of the lender for each action combination.

If we solve that game we arrive at the results pictured in figure 12.5. Obviously, there are multiple equilibria. How does that match with [RS04, Proposition 3] where Rai and Sjöström claim that the game has only unique Nash equilibria? As figure 12.5 indicates: For the shapes considered by Rai and Sjöström the claim is true and the Nash equilibrium is unique. But there exist more shapes in which the equilibrium is not unique! Moreover the east and north shape with equilibria (0R-n|0R-n) and (1R|0R-n) render the game highly vulnerable to collusion, as the borrowers are better of with the (0R-n|0R-n) equilibrium (same amount of joint punishment but less joint repayment).

By the way, this is an illustrating example how the choice of an ability to repay (a random variable, see p. 40) that is too primitive isn't just convenient but can flaw your results. If the modelling rule is 'as simple as possible, as complex as necessary', the two-valued ability to repay used by Rai and Sjöström unfortunately drops out on the 'too simple' side. Realistically seen repayment rates in this game are even worse than in traditional joint liability lending (B.2); either due to collusion, if borrowers can coordinate, or because of coordination failure, if they cannot. The

109 Allowing for partial joint liability – as Rai and Sjöström do by defining $A \equiv \min\{h, 2\bar{R}\}$ – qualitatively does not change the results of the repayment game. Accordingly, the equilibria in cases 1–4 of their proof are the same for all permitted values of h. See [RS04, p. 226-228].

advice that can be given to practitioners is to *not* apply this game in real life under more realistic realisations of the ability to repay.

<table>
<tr><th></th><th></th><th>P2:0R-n</th><th>P2:0R-y</th><th>P2:1R</th><th>P2:2R</th></tr>
<tr><td rowspan="4">Punishment function</td><td>P1:0R-n</td><td>-R | -R</td><td>-e-2R | -P</td><td>-2R | 0</td><td>-P | 0</td></tr>
<tr><td>P1:0R-y</td><td>-P | -e-2R</td><td>-P | -P</td><td>-P | -P</td><td>0 | 0</td></tr>
<tr><td>P1:1R</td><td>0 | -2R</td><td>-P | -P</td><td>0 | 0</td><td>0 | -P</td></tr>
<tr><td>P1:2R</td><td>0 | -P</td><td>0 | 0</td><td>-P | 0</td><td>0 | 0</td></tr>
</table>

<table>
<tr><th></th><th></th><th>P2:0R-n</th><th>P2:0R-y</th><th>P2:1R</th><th>P2:2R</th></tr>
<tr><td rowspan="4">Reward function</td><td>P1:0R-n</td><td>0 | 0</td><td>0 | 0</td><td>0 | 0</td><td>0 | e+R</td></tr>
<tr><td>P1:0R-y</td><td>0 | 0</td><td>0 | 0</td><td>0 | 0</td><td>0 | 0</td></tr>
<tr><td>P1:1R</td><td>0 | 0</td><td>0 | 0</td><td>0 | 0</td><td>e+R | 0</td></tr>
<tr><td>P1:2R</td><td>e+R | 0</td><td>0 | 0</td><td>0 | e+R</td><td>0 | 0</td></tr>
</table>

12.4 Punishment and reward function in [RS04]
These tables reproduce the punishment function and the reward function as defined in [RS04, p. 226]. Compared to the presentation in [RS04] punishments and rewards for both borrowers are combined into a single table. The notation is of course translated into ours according to figure 12.2.

Map of lender payoffs

	A2:0	A2:-R	A2:-2R
A1: 0	0	0 R	2R
A1: -R	0 R	0 2R	2R
A1: -2R	2R	2R	2R

Map of deadweight loss

	A2:0	A2:-R	A2:-2R
A1: 0	-2R	-2R -2R	0
A1: -R	-2R -2R	-2R 0	0
A1: -2R	0	0	0

12.5 Results of the message game in [RS04]
The tables show the map of lender payoff (i.e. repayment received minus rewards paid) and the map of deadweight loss for the message game in [RS04]. For details see B.11.

12.3.2 Improved version

Is there a possibility to improve [RS04] and get rid of the collusion equilibria? If we dive into the details of the traceable solution in B.11, to understand why collusion is attractive, it turns out that we can avoid it by introducing a punishment intensity *M* that is slightly more intense than *R*. Moreover, we can reduce punishment load in the cases where one of the loans is repaid to further improve the game with respect to welfare. The new punishment function is displayed in figure 12.6. The reward function and all the rest remain unchanged.

		P2:0R-n	P2:0R-y	P2:1R	P2:2R
Punishment function (improved)	**P1:0R-n**	-M | -M	-e-2R | -P	-M | 0	-P | 0
	P1:0R-y	-P | -e-2R	-P | -P	-P | -P	0 | 0
	P1:1R	0 | -M	-P | -P	0 | 0	0 | -P
	P1:2R	0 | -P	0 | 0	-P | 0	0 | 0

12.6 Punishment function of the message game in [RS04] – improved
The table pictures an improved version of the punishment function for the message game in [RS04]. Changes compared to Rai and Sjöströms original version are marked with boxes (compare figure 12.4).

This improved version delivers excellent results! As figure 12.7 shows, repayment rates are at its maximum and punishment load is at the level of individual liability with social sanctions (B.8). And this time, all equilibria are unique.

Notice that the improved cross-reporting mechanism (B.12) and individual liability with social sanctions (B.8) provide the same outcomes. So which one should a practitioner apply? Basically the question is if borrowers punish themselves or hand the information over to the lender to let him exert the punishment. Thus, the lender should reflect on his own and the borrowers' incentive tools. If the lender's incentive tools are restricted, choose individual liability and leave the mutual insurance business to the borrowers, as for the cross-reporting mechanism you have to credibly announce punishments of $P > 2R$ in some states; whereas under individual liability $M > R$ is sufficient. Conversely, if the lender has fully fledged incentive tools and social sanctions are weak, choose cross-reporting, get hold of the necessary information and punish accordingly.

Map of lender payoffs

	A2:0	A2:-R	A2:-2R
A1: 0	0	R	2R
A1: -R	R	2R	2R
A1: -2R	2R	2R	2R

Map of deadweight loss

	A2:0	A2:-R	A2:-2R
A1: 0	-2M	-M	0
A1: -R	-M	0	0
A1: -2R	0	0	0

12.7 Results of message game in [RS04] – improved
The tables show the map of lender payoff (i.e. repayment received minus rewards paid) and the map of deadweight loss for the improved version of the message game in [RS04]. For details see B.12. Compared to the original version (see figure 12.5) there are only unique equilibria and the punishment load is reduced.

12.4 Cross-reporting and individual liability

If message games were thought to help distinguish accidental default and strategic default the results so far are slightly disappointing. Can't we build a repayment game with absolutely zero punishment in equilibrium?

In section 3.1.3 I wrote that group lending has many aspects. One aspect is joint liability, another aspect is cross-reporting. So far cross-reporting always appeared in the context of joint liability. But what about cross-reporting *without* joint liability? In this section we will combine cross-messaging with individual liability and – by the way – reduce equilibrium punishment to zero.

Again, the game requires the borrowers to make their repayment and simultaneously send a message. This time the message concerns the realisation of the ability to repay of another borrower.[110] Consequently, the action set of a borrower is {0R-n,0R-y,1R-n,1R-y,2R-n,2R-y}. The possibility to repay 2R is only included for easy comparison with the other repayment games, but no borrower will ever repay 2R without joint liability or mutual insurance.

110 In contrast to the message about the *joint* ability to repay in the games that combine cross-reporting with joint liability

The punishment function applies the following rules in order:

Anti-coordination-failure-rule If both borrowers repay nothing and both borrowers say "y", then both borrowers are punished with $-2M$.

Anti-lying-rule If borrower i is lying, i.e. if borrower i's message does not coincide with borrower j's repayment, then borrower i is punished with $-M$.

Anti-strategic-default-rule If borrower i seemingly strategically defaults, i.e. if borrower i's repayment is 0 although borrower j sent the message "y", then borrower i is punished with $-M$.

In addition the following reward function is applied:

Sincerity-reward-rule If borrower i repays despite being excused, i.e. if borrower i's repayment is $1R$ although borrower j's message is "n", then borrower i is given back his repayment and in addition he is rewarded with $+e$.

Notice that a borrower can be punished and rewarded at the same time, e.g. because of his faulty message concerning his peer and his repaying though excused. The intensities are $R < M < 2R$ and $e > 0$. Figure 12.8 presents the results of this individual liability repayment game with cross-reporting.

Map of lender payoffs

	A2:0	A2:-R	A2:-2R
A1: 0	0	R	R
A1: -R	R	2R	2R
A1: -2R	R	2R	2R

Map of deadweight loss

	A2:0	A2:-R	A2:-2R
A1: 0	0[111]	0	0
A1: -R	0	0	0
A1: -2R	0	0	0

12.8 Results of individual liability with cross-reporting
The tables show the map of lender payoff (i.e. repayment received minus rewards paid) and the map of deadweight loss for the new individual liability with cross-reporting repayment game. For details see B.13. The new game features zero equilibrium punishment load and solely unique equilibria in *every* shape.

111 Without the anti-coordination-failure-rule (0R-n,0R-n) *and* (0R-y,0R-y) would be a Nash equilibrium. Though not a problem for repayment rates the (0R-y,0R-y) equilibrium is welfare decreasing as both borrowers would be punished for lying. The anti-coordination-failure-rules makes (0R-y,0R-y) sufficiently unattractive so that it is never played.

Finally, we actually managed to construct a game with absolutely zero punishment in every equilibrium! Not even the repayment games with social sanctions from the previous chapter are that welfare conserving (compare B.13 with B.8). That is how lending should be: Only strategic defaulters ought to be punished. The game helps creditors to unravel strategic default and accidental defaults. Moreover, all equilibria are indeed unique, there are no ambiguous cases. The repayment quality is of course that of an ordinary individual liability contract.

The game is also relatively collusion proof. In the message game presented in [RS13] it suffices if the borrowers talk to each other to navigate themselves towards the (0R-n|0R-n) equilibrium in shape (A1:-R,A2:-R) (see B.10). Here, in contrast, borrowers would need the ability to threaten each other with social sanctions of $S > e$ to enforce playing the non-equilibrium (0R-n,0R-n) and save repayment. Thus, the higher the promised reward for deviating from 0R-n to 1R-n (sincerity-reward-rule), the more collusion proof this game becomes.

Recall that we need information and incentive tools to enforce repayment (p. 24). In group lending the challenge is to effectively combine the lender's and the borrowers' information and incentive capabilities. One explanation for the trend to switch to individual liability but keep weekly meetings (see [GK14]) are information transfers between lender and borrowers.

Cross-reporting hypothesis. Under the cross-reporting hypothesis presented here information (about another borrower's *ability to repay*) flows from the borrowers to the lender, the punishment is realised via the lender's incentive tools.

Side-contracting hypothesis. Under the side-contracting hypothesis advocated in [RS13] information (about another borrower's *repayment*) flows from the lender to the borrowers, the punishment is realised via the borrowers' incentive tools like socials sanctions.

In a nutshell, the question is: Are the meetings used to *collect* the borrowers' information on each other to apply appropriate bank punishment? Or is the main purpose of public meetings to *give* the borrowers information to enable side-contracts as [RS13, p. 260-262] suggest?

Of course, these questions could most convincingly be answered by empiricists. From a theoretic point of view, however, the information that borrowers gain from public repayment seems insufficient for side-contracting. The flaw in [RS13] is that the variety of shapes considered is incomplete; a borrower either fails and can repay nothing or succeeds and can repay two loans. Thus repaying $1R$ implies success and the ability to also bail a peer out. But what if we incorporate the fairly realistic possibility that the ability to repay of a borrower is just enough to repay his own loan? Then a public repayment of $1R$ does not contain any information about the ability to perform a bail-out – which is the bit of information borrowers would need to close side-contacts. Conclusion: If borrowers don't have the information about each others' ability to repay, public repayment won't give it to them! On these grounds the cross-reporting[112] hypothesis presented here seems more convincing than the side-contracting hypotheses of Rai and Sjöström to explain the adherence to regular meetings under individual liability.[113]

Side-contracting hypothesis	**IL & public repayment** • Information about amount repaid flows from lender to borrowers. • Incentives consist of bank punishment and social sanctions.	**Problematic**, since dependend on an ability to repay with only two possible realisations.
Cross-reporting hypothesis	**IL & cross-reporting** • Information about ability to repay flows from borrowers to the lender. • Incentives consist of bank punishment only.	**Works in general** for all abilities to repay and always provides **maximum efficiency**.

12.9 Reasoning banking meetings by information transfers
Why do banks keep frequent banking meetings whilst switching from joint liability to individual liability? Apart from savings in transaction costs and ideas that regard the meetings as an incentive tool to create public shame, there exist theories that stress the information transfer at banking meetings.

Cross-reporting can get a lender that suffers from hidden information and uses individual liability lending into a position that feels as if he had full information. He can exactly target his punishment at strategic default and avoid punishing accidentally defaulting borrowers.

According to our analysis, the practical advise is that welfare maximising lenders should switch to individual liability with cross-reporting (B.13). Lenders that aim to maximise repayment will certainly prefer joint liability with cross-reporting (B.12) or with adaptive punishment (B.3).[114]

112 Maybe is should better be called cross-auditing, as it is in fact an auditing technique.

113 Nevertheless, even though public repayment does not work as an information tool (to induce side contracts) – the amount repaid does not give the borrowers enough information on their peers' ability for bail-out – it may still serve as an incentive tool (to create public shame).

114 It would be interesting to examine if this dependence of contract choice on a lender's mission can in fact be observed in the real world.

12.5 Summary and future research

After studying a variety of repayment games with cross-reporting, the two most important takeaways of this article are:

1. Given joint liability: Cross-reporting can replace social sanctions.
2. Given individual liability: Cross-reporting can compensate hidden states.

In reality, however, message games may perform less well than expected for a reason pointed out in [BO10, p. 349]. The thing is that borrowers might stop sharing information with each other in order to keep their individual ability to repay secret. Thereby, the mechanism itself would erode the informational advantage of borrowers that it aimed to exploit.

It is also doubtful whether real loan officers of Microfinance institutions implement the mechanism presented here with institutional accuracy. Most likely group meetings will rather produce an informal and diffuse flow of information. Hence, it is more than comprehensible when Armendáriz and Morduch write "... [the mechanism] works well on paper in a specific theoretical context" [AM10, p. 127] and Bhole and Ogden add "However, in reality, microfinance institutions such as Grameen Bank and Spandana do not make use of cross-reporting; especially not in the manner suggested in Rai and Sjostrom (2004)." [BO10, p. 349]

It would therefore be interesting if we could come up with some theoretical concept that models information and its exchange less explicitly and somehow more unspecific. So far messages have a distinct sender and receiver and only two possible values ("y" or "n"). But of course human communication is much more vague and fuzzy. A research project that proceeds in that direction would certainly be a conceptual challenge.

It would also be of interest to extend the games in this article to larger groups of borrowers. However, be aware of several conceptual hurdles! It is always the same story: In reality group lending with larger groups is quite common, but the theory jumps to a whole new level of complexity for group sizes that exceed $n = 2$. What would a message mean? Do we need to make the message string more complicated, e.g. "n,y" to give a borrower the chance to report on each of his peers? What are the punishments if the messages about a particular borrower do not coincide? Maybe knowledge from implementation theory will help. Can the mechanism of Maskin (1999) mentioned in [MS02, Theorem 2, p. 249] with its modulo game in rule 3 somehow be applied?

In general implementation theory seems to be promising and I doubt that all of its insights have already been applied to lending theory. In implementation theory – a branch of game theory – the outcomes that are desirable are specified for each state of the world. The technical term for the specification of desired outcomes is called a *social choice rule* and if a mechanism produces equilibrium outcomes that coincide with such a rule, then we say this rule is *implemented* by the mechanism. Now the goal of implementation theory is not to just find such mechanisms in certain cases but, as Maskin and Sjöström emphasise, to characterise "the full class of implementable social choice rules." [MS02, p. 240]

In the theory of repayment games we face exactly this type of problem. We have a desired outcome (i.e. that borrowers repay if the are able to) and the question is if there exists a mechanism that implements this outcome in each possible state of the world. So implementation theory may bear further fruits in lending theory. A rigorous presentation of the subject is provided by the well-written text of Maskin and Sjöström "Implementation Theory" (see [MS02]).

13 Outlook: Modularisation and standards improve research quality

Abstract This final chapter presents a vision for the future of credit contract theory and makes three proposals to improve its quality and applicability.

During the time I spent as a PhD student in the economic research community, I discovered that amongst economists there seems to be no common ground on the raison d'être of models. Some say it suffices if some specific models demonstrate basic principles to train market participants and politicians engaged in economic policy, who then take decisions with their expert eye. Such research is based upon economic fables that use simple assumptions[115] to illustrate certain aspects of lending theory.

In my natural scientist view, however, the reason why we put effort and time in building theoretic model is to make predictions! Despite all the haziness in social sciences, compared to natural sciences, I believe that we should strive to model the behaviour of economic agents as well as possible to gain high-quality projections. Showcase models will not help us as their applicability remains questionable – they may even be misleading in more complex situations (see below). To equip our decision makers with precise predictions we need models that are based on reasonably realistic assumptions.

The modular approach presented in this dissertation, which decomposes lending theory into is individual aspects and links them by adequate interfaces, is the tool to build such more realistic models without increasing overall complexity to a level that couldn't be handled anymore. The following three suggestions may therefore serve as a guideline to improve research usability and applicability.

115 E.g. Bernoulli distributions for real-world processes as complex as the earnings of an investment project.

C. Prem, *The Theory of Credit Contracts*,
https://doi.org/10.1007/978-3-658-29362-8_13

13.1 Suggestion 1: Research sub-problems separately

Given we want to build a monolithic model that includes the ability to repay, a certain repayment game, utility functions for the borrowers and a market situation. How many choices can we make? Let's be very conservative and say there are only four possibilities to model the ability to repay, namely a joint uniform distribution, a joint distribution with negative or positive correlation and a joint distribution with tail dependence. If we just consider the repayment games studied in this dissertation we can choose amongst at least 10. Let's say borrowers can be either risk-neutral, risk-seeking or risk-averse. And say there are only two choices for the market structure, namely a monopoly lender or perfect competition. This amounts to $4 \cdot 10 \cdot 3 \cdot 2 = 240$ possible monolithic models. If we want to pairwise compare the outcomes it will be $(240 \cdot 239)/2 = 28680$ comparisons. And we haven't talked about even more choices like loan size, group size, duration or monitoring strategies.

What might delight the scientific journal industry won't be useful to the public, which pays most researchers' wages. Unfortunately, it seems to be a trend to publish as much as possible, instead of publishing few well-thought-out general results. It remains doubtful whether such research strategy will deepen our understanding of credit contracts. Instead, it will become impossible to keep track of the results and compare them – especially if the notation changes from paper to paper. It may even become impossible to just read all of these publications. Altogether: Building monolithic models will make us drown in a huge salad bowl of lending models.

Modular research, in contrast, creates clarity by avoiding redundancy. It promotes cooperation instead of research quantity competitions. For instance, take the repayment games studied in chapter 10-12. The mechanisms B.3, B.8 and B.12 enforce the same repayment function. Thus, if we calculate expected repayment for one repayment function, we cover three repayment games. We research the *impact* of a repayment function (expected repayment, risk-shift effects, ...) separately from its *implementation* (the repayment game in the payback-enforcement module). As explained in chapter 2 the trick is to put details in a black box and works with interfaces between the different aspects of credit contracts (in our example the interface is the map of repayment). Altogether, modular research is much more efficient.

It is also more insightful as the interplay between different aspects comes forward quite clearly. The practice of decomposing a big problem into reasonably isolated subproblems and designing adequate interfaces brings out interdependencies much better than a monolithic model could ever do. And for applications, the results from individual modules can easily be assembled to fit the actual real-world situation – like with toy blocks.

13.2 Suggestion 2: Research modules comprehensively

In the course of this dissertation we encountered instances where unrealistically simple assumptions led to misleading results. Number one was the positive assortative matching result of [Gha99] which can break down under the realistic assumption of correlated abilities to repay (see [KL12]). Number two was screening via collateral in [Bes85] which is impaired by the realistic assumption of risk-averse borrowers (see 7.5). Number three was the side-contracting induced by public repayment in [RS13] which does not work if we add the realistic possibility that a borrower may end up *between* being extraordinarily successful or entirely bankrupt (see p. 189).

The reason for such assumptions may be the wish to keep a model simple. But such models are unrealistic special cases that only shed light onto corners of the space of possibilities. Instead we need to understand how certain contracts perform under a variety of different assumptions before we give advice to decision makes in the real world. We need to research lending theory comprehensively!

Concerning the payback-enforcement module a first start has been made in this dissertation. We have analysed a multitude of known and new repayment games. We did not simplify by ignoring undesired equilibria (as [BC95] does), we did not simplify by neglecting shapes (e.g. as in [BC95] or [RS04]), we varied available punishment intensity, we considered individual liability and joint liability, we calculated repayment with and without social sanctions. And still, this study of the payback-enforcement module is only the beginning. We could for example combine message games and social sanctions in one game. Another open question: What happens if borrowers close a joint liability contract but in the end do not get to know each other's realisations' of the ability to repay? Certainly, cross-reporting and precisely targeted social sanctioning will fail. But moreover, the borrowers do not even know which shape they play and instead have to form beliefs. Can group lending work in an urban environment, between unrelated people? Further research should also combine joint liability with collateral, it should include asymmetric games (with just unilateral liability instead of joint liability) and most importantly it should analyse more realistic continuous repayment games (with continuous action sets) because, in reality, the lender will retrieve as much as possible, even if it is a repayment of $0.19850607R$. For continuous games things certainly get more complex. Action sets are infinite, punishment functions are not discrete any more, and to analyse the equilibrium of such games new concepts like the ε *Nash equilibrium solution* may be required (see [BO99, Section 4.2]). But: Only by researching the modules comprehensively we will gain a realistic impression and broad knowledge about lending theory. Comprehensive research will reveal which economic fables are representative and which are just odd special cases.

13.3 Suggestion 3: Present results uniformly

Finally, to keep track of our findings and render theoretical research useful for practitioners, we need a way that clearly arranges our results. For the payback-enforcement module a suggestion is made in figure 13.2.

From the user's viewpoint the modular research strategy should feel like a model kit where he or she can individually assemble the blocks that fit his or her situation. Is he a monopolist? Which incentive tools can she use? Are the borrowers risk-averse? Do they have information about each other's ability to repay? The situations are so different that only a model kit can cover them all. Rather than arguing whether group lending or individual lending is superior with fully-fledged monolithic models, future lending theory should provide individually tailored advice for specific situations. Let's create a toolbox of theoretically proven and empirically verified lending techniques!

In chapter 1 we have seen how important credits are. Deferred reciprocity agreements are omnipresent. Improving this tool means improving the way we distribute resources and cooperate, it means perfecting an essential corner stone of our living together. It is definitely worth the effort!

Description	**EqRepay**	**EqPunLoad**
IL = individual liability JL = joint liability wSS = with social sanctions wCR = with cross-reporting	□ Joint repayment 0R ◢ Joint repayment 1R ■ Joint repayment 2R ! Multiple equilibria ? Ambiguous case	□ Joint punishment 0 ◢ Joint punishment $1P_M$ ■ Joint punishment $2P_M$ ◢ Joint punishment $1P_H$ ■ Joint punishment $2P_H$! Multiple equilibria ? Ambiguous case
Information ✉		**Incentives ↯**
"-" = no prerequisites "full" = full information on the realisations of the abilities to repay		$R < S_M < 2R$ $R < P_M < 2R$ $2R < P_H$ $0 < e$

13.1 Legend for the catalogue of payback-enforcement schemes
Explanation of the symbols used in figure 13.2.

SCHEME		PREREQUISITES				PERFORMANCE	
		Creditor		Debtor		Map of	
Description	Ref	✉	↯	✉	↯	EqRepay	EqPunLoad
IL	B.1	–	$-P_M$	–	–		
JL high	B.2	–	$-P_H$	full	–		
JL medium	B.5	–	$-P_M$	full	–		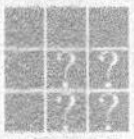
JL adaptive	B.3	–	$-P_H$ $-P_M$	full	–		
IL wSS	B.8	–	$-P_M$	full	$-S_M$		
JL high wSS	B.4	–	$-P_H$	full	$-S_M$		
JL medium wSS	B.6	–	$-P_M$	full	$-S_M$		
IL wCR	B.13	–	$-P_M$ $+e$	full	–		
JL wCR RS04 improved	B.12	–	$-P_H$ $-P_M$ $+e$	full	–		
JL wCR RS13 corrected	B.10	–	$-P_H$ $+e$	full	–		

13.2 Catalogue of payback-enforcement schemes
This figures lists the (most important) repayment games treated in this dissertation. At the same time it demonstrates what a well-structured presentation of our knowledge about the payback-enforcement module could look like. For the legend see figure 13.1.

Part III

Appendix

C. Prem, *The Theory of Credit Contracts*,
https://doi.org/10.1007/978-3-658-29362-8

A Software code and documentation

A.1 General concepts

This section introduces general concepts used throughout the algorithm. It is necessary to understand them in order to understand the subsequent documentation.

A.1.1 Information Analysis

First of all we have a number of players `nrOfPlayers`. Note that the number of players is a constant within a game. But since we want to write a piece of software that works for an arbitrary numbers of players, it is a symbol in our code. Each player has a unique `playerKey`.

A single action of a player is a tuple with three components:

`actionName`	can be any description of the action
`actionRepayment`	the repayment associated with this action in the repayment game
`actionMessage`	a potential message sent to the principal (or empty String)

A player's `actionSet` is an ordered list of actions. An action can be addressed unequivocally by its `actionSetKey`, which is its position in the ordered `actionSet`.

An `ability` of a player specifies a upper bound for a player's `actionRepayment`.

Punishments reflect the influence on a borrower's payoff exerted by the lender. A `punishmentRule` consists of two pieces of information:

`punishmentValue`	the value by which a borrower's payoff is reduced by the punishment (the intensity of the punishment)
`punishmentCondition`	a formal condition under which a punishment is to be executed

An ordered list of `punishmentRules` is called a `punishmentFunction`. The order reflects the hierarchy of the rules; the first rule whose `punishmentCondition` yields true is applied.

C. Prem, *The Theory of Credit Contracts*,
https://doi.org/10.1007/978-3-658-29362-8

Sanctions reflect the influence on a borrower's payoff exerted by other borrowers. A `sanctionRule` consists of two pieces of information:

`sanctionValue`	the value by which a borrower's payoff is reduced by the sanction (the intensity of the sanction)
`sanctionCondition`	a formal condition under which a social sanction is to be executed

An ordered list of `sanctionRules` is called a `sanctionFunction`. The order reflects the hierarchy of the rules; the first rule whose `sanctionCondition` yields true is applied.

`Payoff`	a number describing the payoff that a player receives if a certain action combination is played
`Feasibility`	a boolean value represting the information if a certain action combination can be played, i.e. is feasible

A.1.2 Template types

Furthermore we have some sort of abstract compound data types that refer to different arrangement of the more basic data types introduces above. I call these 'abstract' because they are like abstract compound data but the entries of which the compound data type consists is not specified yet.

`xxxSet`	a list of all xxx of a particular player (e.g. actionSet)
`xxxCombination`	one xxx for all players of the game (e.g. actionCombination)
`xxxMatrix`	one xxx for all actionCombinations (e.g. payoffMatrix)
`xxxTos`	one xxx for all abilityCombinations (e.g. feasibilityMatrixTos), read "tabular of shapes"
`xxxKey`	the position of an entry in the list xxx (e.g. actionSetKey, actionCombinationKey)

A.1.3 Sign convention

Repayments, punishments, sanctions, payoffs are denoted from the perspective of a borrower. That is cash flows reducing a borrower's wealth have a negative sign (flowing away from the borrower) and vice versa.

A.1.4 Global variables

These variables are *global* variables. They are specified in the game definition block.

`n` number of players

`actionSet[i]`
actionSet of player *i*

`punishmentFunction[abilityCombination, actionCombination, playerKey]`
punishment strategy of the lender (loss-flow: borrower → deadweight-loss)

`sanctionFunction[abilityCombination, actionCombination, playerKey]`
rules for social sanctions (loss-flow: borrower → deadweight-loss)

`rewardFunction[abilityCombination, actionCombination, playerKey]`
reward strategy of the lender (transfer-flow: lender → borrower)

`payoffLayers[abilityCombination, actionCombination, playerKey]`
combination of all of the borrower's payoff layers (apart from repayment)

`$Assumptions`
assumptions about the symbols for repayment, punishment, sanctions, rewards

A.1.5 Further Comments

The algorithm is implemented using Mathematica 8.0 syntax only.

The algorithm is realized using the functional programming paradigm. By this approach we avoid introducing many additional variables that would be necessary under the procedural programming paradigm in order to check dimensions and control loops. The functional programming paradigm allows to not bother about dimensions at all, such that the algorithm works for games of arbitrary size (number of players, number of actions) without further modification. Only the depth of arrays becomes relevant from time to time to ensure that the higher-order functions `Map[]` and `MapIndexed[]` work at the right level of a list of lists of lists ....

Notice that some parts of a repayment game naturally can be captured by maps. E.g. we could have functions that assign to each action from the actionSet of a player the repayment amount associated with each action. Likewise for messages associated with a certain action. However, in the following code, I make use of the fact, that each map can be seen as a relation: In technical terms a function can be described as a relation (a set of ordered pairs) that is left-total and right-unique. So instead of defining a repayment function or a message function I represent the same information by defining actions as tuples consisting of (`actionName`, `actionRepayment`, `actionMessage`).

I adopt the following naming conventions:

- Functions commence with `fn...` and procede in upper camel case.
- Functions returning a single atomic boolean value end with `Q` (according to the Mathematica convention).
- Functions that are meant to be applied to individual entries of lists and spring into action via mapping functions like `Map[]` or `MapIndexed[]` are called `fnRobots` (see functional programming paradigm approach, explained above).

Functions are ordered from auxilary functions to combining functions, as Mathematica has to read the auxilary functions first before other functions can call them. Unfortunately this puts the most important functions last.

A.2 Code block: Game construction

A.2.1 Problem Statement

Write a piece of software that helps to define a repayment game (which in turn can be solved by the function `fnSolveGame`). Concretely the program should take as input: the number of players, the actions of each player, payoff layers and global assumptions. The output shall be – for each possible repayment ability-combination – a list of payoff matrices, one for each player, produced from the given rules and assumptions. Furthermore another output should be a feasbility matrix, i.e. the algorithm should automatically know what different ability-combinations can arise and construct the feasibility matrix for each situation.

A.2.2 Brief description of the algorithm

The code is split into several parts.

1. The first part provides auxilary functions working on actions. These function extract components of actions. They are useful to define punishmentConditions or sanctionConditions. They also help to display appropriate headings in TabularForm or MatrixForm of xxxMatrices.
2. The second part constructs an abilitySet for each player. The data type for abilities is simpler that that for actions as they don't have names and messages that come with it.
3. This part deals with xxxKey to xxx transformations. The xxxKeys are important to position entries in xxxMatrices at their correct position (e.g. the payoff under a certain actionCombination).
4. These functions basically measure the size of the game and define templates on which second order functions like Map[] oder MapIndexed[] can operate.
5. Here we construct the payoffMatrix for each player. The function payoffMatrix produces the payoffMatrix after which the function game produces the payoffMatrices of all players and combines them to a payoffMatrixCombination, i.e. the game.
6. The last part finally constructs the feasibilityMatrix that is characteristic for each abilityKeyCombination. What we do here is define a predicate fnFeasibleForPlayerQ which can test whether a certain actionKey is feasible given a certain abilityKey. Next we use this predicate to do something, namely fnDetermineInfeasibleActionKeys collects all infeasible actionKeys of a player for a given abilityKey. Finally, we need to use this list of infeasible actionKeys to find the positions in the feasibilityMatrix that have to be changed to False; the function feasibilityMatrix does this replacement for all players and returns the ready cooked feasibilityMatrix.

A.2.3 Source Code

```
(* ================== *)
(* DATA DEFINITIONS *)

(* -------- *)
(* PLAYERS *)

(* nrOfPlayers *)
(* DEFINITION         is positive Integer *)
(* INTERPRETATION     the total number of players *)
(* EXAMPLES           n=2; n=5; *)
(* TEMPLATE           fnForN[n_]:= ... n ; *)
(* RULES USED         atomic non-distinct: positive Integer *)

(* playerKey *)
(* DEFINITION         is Integer [1,nrOfPlayers] *)
(* INTERPRETATION     the number of a certain player; also represents the order in which players move in sequential games *)
(* EXAMPLES           playerNrMax=1; playerNrMoritz=2; *)
(* TEMPLATE           fnForPlayerNr[playerKey_]:= ... playerKey ; *)
(* RULES USED         atomic non-distinct: Integer [1,nrOfPlayers] *)

(* ---------------------- *)
(* ACTIONS & ABILITIES *)

(* actionName *)
(* DEFINITION         is String *)
(* INTERPRETATION     the name of an action (used as heading in payoffMatrix) *)
(* EXAMPLE            actionName1="repay both", actionName2="default-n", actionName3="-2R&n" *)
(* TEMPLATE           fnForActionName[actionName_]:= ... actionName; *)
(* RULES USED         atomic non-distinct: String *)

(* actionRepayment *)
(* DEFINITION         is negative Real number, or symbol representing a Real number *)
(* INTERPRETATION     the repayment associated with a certain action, i.e. the amount by which a certain repayment reduces the payoff of a borrower *)
(* EXAMPLES           actionRepayment1=-2R, actionRepayment2=-1R, actionRepayment3=0 *)
(* TEMPLATE           fnForActionRepayment[actionRepayment_]:= ... actionRepayment; *)
(* RULES USED         atomic non-distinct: Number *)

(* actionMessage *)
(* DEFINITION         is String *)
(* INTERPRETATION     a messages associated with a certain action, i.e. a message that is sent to the lender *)
(* EXAMPLES           actionMessage1="n", actionMessage2="no", actionMessage3="yes", actionMessage4="" *)
(* TEMPLATE           fnForActionMessage[actionMessage_]:= ... actionMessage; *)
(* RULES USED         atomic non-distinct: String *)

(* action *)
(* DEFINITION         is list of {actionName,actionRepayment,actionMessage} *)
(* INTERPRETATION     an action a player can choose to play in the repayment game *)
(* EXAMPLE            action1={"repay both",-2R,""}, action2={"default",0,"yes"} *)
(* TEMPLATE           fnForAction[action_]:= ... action[[1]], action[[2]], action[[3]]; *)
(* RULES USED         compound: 3 fields , references: actionName is actionName, actionRepayment is actionRepayment, actionMessage is actionMessage *)

```

```
(* nrOfActions *)
(* DEFINITION           is positive Integer *)
(* INTERPRETATION       the total number of a player's actions, i.e. the length of his actionSet *)
(* EXAMPLES             n=2; n=5; *)
(* TEMPLATE             fnForN[n_]:= ... n ; *)
(* RULES USED           atomic non-distinct: positive Integer *)

(* actionSet *)
(* DEFINITION           is list of actions *)
(* INTERPRETATION       all actions a player can ever play *)
(* EXAMPLE              actionSet1={{"opt1",0,"n"},{"opt2",0,"y"},{"opt3",-R,"y"},{"opt4",-2R,"y"}} *)
(* TEMPLATE             fnForActionSet[ actionSet_]:= ... actionSet[[i]] *)
(* RULES USED           compound: as many fields as a player has actions, references: action *)

(* actionSetKey *)
(* DEFINITION           is Integer *)
(* INTERPRETATION       the position of an action in a player's actionSet *)
(* EXAMPLES             actionSetKeyRepay=1; actionSetKeyDefault=2 *)
(* TEMPLATE             fnForActionSetKey[ actionSetKey_]:= ... actionSetKey ; *)
(* RULES USED           atomic non-distinct: Integer *)

(* actionSetKeyCombination *)
(* DEFINITION           is List of positive Integers *)
(* INTERPRETATION       an action combination in the game, the first entry denotes the actionSetKey of the first player etc. *)
(* EXAMPLES             actionSetKeyCombinationA={1,1,1}; actionSetKeyCombinationB={1,1,2} *)
(* TEMPLATE             fnForActionSetKeyCombination[actionSetKeyCombination_]:= ... actionSetKeyCombination [[...]] ; *)
(* RULES USED           compound: nrOfPlayers fields *)

(* ability *)
(* DEFINITION           is negative Real number, or symbol representing a negative Real number *)
(* INTERPRETATION       the maximum repayment a player can pay in a certain situation *)
(* EXAMPLE              ability1=-2R, ability2=-1R, ability3=-10, ability4=0 *)
(* TEMPLATE             fnForAbility[ ability_]:= ... ability ; *)
(* RULES USED           atomic non-distinct: Real number *)

(* abilitySet *)
(* DEFINITION           is list of abilities *)
(* INTERPRETATION       all relevant abilities that can happen to a player *)
(* EXAMPLE              abilitySet1={0,-1R,-2R}, abilitySet2={0,-2R} *)
(* TEMPLATE             fnForAbilitySet[ abilitySet_]:= ... abilitySet[[i]] *)
(* RULES USED           compound: as many fields as a player has abilities, references: ability *)

(* nrOfAbilities *)
(* DEFINITION           is positive Integer *)
(* INTERPRETATION       the total number of a player's abilities, i.e. the length of his abilitySet *)
(* EXAMPLES             n=2; n=5; *)
(* TEMPLATE             fnForN[n_]:= ... n ; *)
(* RULES USED           atomic non-distinct: positive Integer *)

(* abilitySetKey *)
(* DEFINITION           is Integer *)
(* INTERPRETATION       the position of an ability in a player's abilitySet *)
(* EXAMPLES             abilitySetKeyRepay=1; abilitySetKeyDefault=2 *)
```

```
(* TEMPLATE            fnForAbilitySetKey [ abilitySetKey_ ]:= ...  abilitySetKey  ; *)
(* RULES USED          atomic non-distinct: Integer *)

(* abilitySetKeyCombination *)
(* DEFINITION          is List of positive Integers *)
(* INTERPRETATION      an ability combination in the game, the first entry denotes the abilitySetKey of the first player etc. *)
(* EXAMPLES            abilitySetKeyCombinationA={1,1,1}; abilitySetKeyCombinationB={1,1,2} *)
(* TEMPLATE            fnForAbilitySetKeyCombination[ abilitySetKeyCombination_ ]:= ...  abilitySetKeyCombination [[...]]  ; *)
(* RULES USED          compound: nrOfPlayers fields *)


(* ----------------------- *)
(* PUNISHMENTS & SANCTIONS *)

(* punishmentValue *)
(* DEFINITION          is (negative / positive ) Real *)
(* INTERPRETATION      the amount by which a player's payoff is decreased/increased by the lender *)
(* EXAMPLE             punishmentValue1=-p, punishmentValue2=+e *)
(* TEMPLATE            fnForPunishmentValue[punishmentValue_]:= ...  punishmentValue *)
(* RULES USED          atomic non-distinct: Real number *)

(* sanctionValue *)
(* INTERPRETATION      the amount by which a player's payoff is decreased/increased by other borrowers *)
(*                     likewise *)

(* punishmentCondition *)
(* DEFINITION          is a predicate (i.e. a function that returns a Boolean) with signature {abilitySetKeyCombination,actionSetKeyCombination,playerKey} --> Boolean *)
(* INTERPRETATION      the condition under that decides whether a certain punishment is applicable, i.e. under which circumstances the punishment gets executed *)
(* EXAMPLE             Plus@@(fnGetActionRepayment/@actionCombination)<=-n R, Union[fnGetActionMessage/@actionCombination]=={"y"} *)
(* TEMPLATE            fnForPunishmentCondition[punishmentCondition_]:= If[punishmentCondition, ...] *)
(* RULES USED          the data type here should be 'function', as the information here is captured by a function *)

(* sanctionCondition *)
(*                     likewise *)

(* punishmentRule *)
(* DEFINITION          is a list of {punishmentValue, punishmentRule} *)
(* INTERPRETATION      the amount of a punishment and the condition under which the punishment gets executed *)
(* EXAMPLE             punishmentRule={0,Union[fnGetActionMessage/@actionCombination]=={"y"}} *)
(* TEMPLATE            fnForPunishmentRule[punishmentRule_]:= ...  If[ punishmentRule[[2]],punishmentRule[[1]] ] *)
(* RULES USED          compound: 2 fields, references: punishmentValue is punishmentValue, punishmentCondition is punishmentCondition *)

(* sanctionRule *)
(*                     likewise *)

(* punishmentFunction *)
(* DEFINITION          an (ordered) list of punishmentRules *)
(* INTERPRETATION      specification of the complete punishment regimen, the first punishmentRule whose punishmentCondition yields true is applied *)
(* EXAMPLE             punishmentFunction1={{0, Union[fnGetActionMessage/@actionCombination]=={"y"}]} (*Rule1*),{+2R+e, Plus@@(fnGetActionRepayment/@actionCombination)<=-n R}}(*Rule2*)} *)
(* TEMPLATE            fnForPunishmentFunction[punishmentFunction_]:= ...  Piecewise[punishmentFunction]; *)
(* RULES USED          compound: arbitrary many fields, reference: punishmentRule *)

(* sanctionFunction *)
```

```
(*                      likewise  *)


(* -------------------- *)
(* PAYOFF & FEASBILITY *)

(* payoff *)
(* DEFINITION          is Real (or a Mathematica expression representing a Real number) *)
(* INTERPRETATION      the payoff a player receives if a certain action combination is played *)
(* EXAMPLES            payoffCombination111Player1=7; payoffCombination111Player2=-2R-p *)
(* TEMPLATE            fnForPayoff[ payoff_ ]:= ... payoff ; *)
(* RULES USED          atomic non-distinct: Real number *)

(* feasibility *)
(* DEFINITION          is Boolean *)
(* INTERPRETATION      true if an action combination is feasible , false otherwise *)
(* EXAMPLES            feasibilityCombination111 =True; pfeasibilityCombination112 =False *)
(* TEMPLATE            fnForFeasibility [ feasibility_ ]:= ... feasbility ; *)
(* RULES USED          atomic non-distinct: Boolean *)


(* --------------- *)
(* TEMPLATE TYPES *)

(* xxxSet *)
(* DEFINITION          is List of arbitrary size with entries xxx (of arbitrary type) *)
(* INTERPRETATION      all xxx of a particular player *)
(* EXAMPLES            actionSet , repaymentSet, messageSet, abilitySet *)
(* TEMPLATE            fnForxxxSet[ xxxSet_]:= ... xxxSet[[ i ]]; *)
(* RULES USED          compound: arbitrary many fields *)

(* xxxCombination *)
(* DEFINITION          is a List of length nrOfPlayers (i.e. one entry for each player) with entries xxx (of arbitrary type) *)
(* INTERPRETATION      first position =xxx of first player, second position =xxx of second player, etc. *)
(* EXAMPLES            actionCombination, actionSetKeyCombination, abilityCombination , payoffCombination, messageCombination, nrOfActionsCombination, payoffMatrixCombination *)
(* TEMPLATE            fnForxxxCombination[xxxCombination_]:= ... xxxCombination[[playerKey]]; *)
(* RULES USED          compound: nrOfPlayer many fields *)

(* xxxMatrix *)
(* DEFINITION          is List of Lists of Lists ... of dimensions nrOfActionsCombination, with content xxx (of arbitrary type) *)
(* INTERPRETATION      one entry for each actionCombination, a position is xxx belonging to this actionCombination *)
(* EXAMPLES            payoffMatrix , feasibilityMatrix *)
(* TEMPLATE            fnForMatrix[ xxxMatrix_]:= ... xxxMatrix[[actionCombination]] ; *)
(* RULES USED          compound: the product of nrOfActionsCombination many fields (the dimensions of the game) *)

(* xxxTos *)
(* DEFINITION          is List of Lists of Lists ... of dimensions nrOfAbilitiesCombination , with content xxx (of arbitrary type) *)
(* INTERPRETATION      one entry for each abilityCombination (each shape of the game), a position is xxx belonging to this abilityCombination *)
(* EXAMPLES            payoffMatrixTos , feasibilityMatrixTos *)
(* TEMPLATE            fnForxxxTos[xxxTos]:= ... xxxTos[[ abilityCombination ]] ; *)
(* RULES USED          compound: the product of nrOfAbilitesCombination many fields *)

(* xxxKey *)
```

```
(* DEFINTIION          is positive Integer *)
(* INTERPRETATION  the position of an entry in the list xxx *)
(* EXAMPLES            actionSetKey, abilitySetKey *)
(* TEMPLATE            fnForxxxKey[xxxKey]:= ... xxx[[xxxKey]] ; *)
(* RULES USED          atomic non-distinct: positive Integer *)

(* nrOfxxx *)
(* DEFINTIION          is positive Integer *)
(* INTERPRETATION  the number of xxx *)
(* EXAMPLES            nrOfPlayers, nrOfActions *)
(* TEMPLATE            fnForNrOfxxx[nrOfxxx]:= ... nrOfxxx; *)
(* RULES USED          atomic non-distinct: positive Integer *)

(* game = payoffMatrixCombination *)
(* DEFINITION          is List of payoffMatrices *)
(* INTERPRETATION  the first payoffMatrix describes the payoffs of the first player ... etc. *)
(* EXAMPLES            game1={2 3
5 1

,6 2.1
13 2

}; *)
(* TEMPLATE            fnForGame[game_]:=... game [[...]] (*payoffMatrix*); *)
(* RULES USED          compound: nrOfPlayers fields ; xxxMatrix; payoff; *)



(* ========== *)
(* FUNCTIONS *)


(* ---------------1--------------- *)
(* Auxilary functions for actions *)

(* fnGetAction ... *)
(* SIGNATURE         [ action_] --> actionName/actionRepayment/actionMessage *)
(* PURPOSE             extract name/repayment/message of an action *)
(* TESTS *)
fnGetActionName[{"repay",-R,"no"}]==="repay";
(* FUNCTION BODIES *)
fnGetActionName[action_]:=action [[1]]; (* select actionName of an action *)
fnGetActionRepayment[action_]:=action [[2]] ;(* select actionRepayment of an action *)
fnGetActionMessage[action_]:=action [[3]] ;(* select actionMessage of an action *)
(* USAGE:
1) these functions can be applied to all lists of actions entrywise by using /@ (e.g. actionSets or actionCombinations are lists of actions).
 2) (fn .../ @ actionCombination)[[playerKey]] transforms the whole actionCombination using the function fn ... and then picks out playerNr's part.
        It is equivalent to fn ... @(actionCombination[[playerKey]]) which picks playerNr's action first and then transforms it and is thus less time-consuming.*)

(* --------------2--------------- *)
(* Construction of abilitySets *)

(* abilitySet *)
```

```
(* SIGNATURE          [playerKey_] --> abilitySet *)
(* PURPOSE            construction of a player's abilitySet , i.e. all relevant abilities considering his repaymentSet *)
(* TESTS *)
(* would interfere with the global variables actionSets [i] *)
(* FUNCTION BODY *)
abilitySet [playerKey_]:= DeleteDuplicates [fnGetActionRepayment/@actionSet[playerKey] ]
(* COMMENTS
Since it is a function it should actually be called fnAbilitySet . In analogy to the data type actionSets I nevertheless dropped the fn ... prefix . Regard each ability as an additional nature-player with
     actionSet = abilitySet . However, the player does not optimize his behaviour, has no payoffMatrix and is instead steered by chance. *)

(* -----------------3---------------------- *)
(* Key to Action/ Ability transformations *)

(* fnKeyToAction / fnKeyToAbility respectively *)
(* SIGNATURE          [playerKey_, actionKey_] --> action *)
(* PURPOSE            get action belonging to a particular actionKey *)
(* TESTS *)
(* would interfere with global variables actionSet [i] *)
(* FUNCTION BODY *)
fnKeyToAction[playerKey_,actionKey_]:= actionSet [playerKey ][[ actionKey ]]; (* actionKey to action *)
fnKeyToAbility [playerKey_, abilityKey_ ]:= abilitySet [playerKey ][[ abilityKey ]]; (* abilityKey to ability *)

(* fnKeyToActionCombination / fnKeyToAbilityCombination respectively *)
(* SIGNATURE          [actionKeyCombination_] --> actionCombination *)
(* PURPOSE            transform an actionKeyCombination into the corresponding actionCombination *)
(* TESTS *)
(* would interfere with the global variables actionSets [i] *)
(* FUNCTION BODY *)
fnKeyToActionCombination[actionKeyCombination_]:=MapIndexed[fnKeyToAction[Sequence@@#2,#1]&,actionKeyCombination,{1}]; (* actionKeyCombination to actionCombination *)
fnKeyToAbilityCombination[abilityKeyCombination_]:=MapIndexed[fnKeyToAbility[Sequence@@#2,#1]&,abilityKeyCombination,{1}]; (* abilityKeyCombination to abilityCombination *)

(* --------------4------------------ *)
(* Measuring the size of the game *)

(* nrOfActionsCombination / nrOfAbilitiesCombination respectivels *)
(* SIGNATURE          [ ] --> nrOfActionsCombination *)
(* PURPOSE            measure the number of actions of each player *)
(* TESTS *)
(* would interfere with the global variables actionSets [i] *)
(* FUNCTION BODY *)
nrOfActionsCombination:=Map[Length[actionSet[#]]&,Range[n],{1}] (* measuring the number of actions of each player, i.e. dimensions of xxxMatrix *)
nrOfAbilitiesCombination :=Map[Length[abilitySet[#]]&,Range[n],{1}]; (* measuring the number of abilities of each player, i.e. dimensions of xxxTos *)

(* emptyMatrix / emptyTos respectively *)
(* SIGNATURE          [ ] --> xxxMatrix *)
(* PURPOSE            construct a template needed for payoffMatrix construction *)
(* TESTS *)
(* ... *)
(* FUNCTION BODY *)
emptyMatrix:=Array[List,nrOfActionsCombination]; (* one entry for all possible actionCombinations, captures dimensions of xxxMatrix, i.e. size of the game *)
emptyTos:=Array[List,nrOfAbilitiesCombination ]; (* one entry for all possible abilityCombinations, captures dimensions of xxxTos, i.e. all possible shapes of the repayment game *)
(* emptyCombination:=Range[n]; So simply write Range[n] to save variable names. *)
```

```
(* ----------5--------- *)
(* game construction *)

(* payoff *)
(* SIGNATURE           [abilityKeyCombination_, actionKeyCombination_playerKey_] --> payoff *)
(* PURPOSE             calculate a player's payoff in a particular situation *)
(* TESTS *)
(* depend on global variables punishmentFunction, sanctionFunction, actionSets *)
(* FUNCTION BODY *)
fnPayoff[abilityKeyCombination_, actionKeyCombination_, playerKey_]:=Module[
{abilityCombination, actionCombination, payoff},
abilityCombination=fnKeyToAbilityCombination@abilityKeyCombination; (* get from abilityKeys to abilities, since punishmentFunction and sanctionFunction need these *)
actionCombination=fnKeyToActionCombination@actionKeyCombination; (* get from actionKeys to actions, since punishmentFunction and sanctionFunction need these *)
payoff=fnGetActionRepayment@(actionCombination[[playerKey]])
+ payoffLayers[abilityCombination, actionCombination, playerKey]; (* calculate payoff *)
Return[payoff]; (* return calculated payoff *)
];

(* payoffMatrix *)
(* SIGNATURE           [abilityKeyCombination_, playerKey_] --> payoffMatrix *)
(* PURPOSE             robot that constructs the payoff-matrix for a certain player in a certain shape (i.e. for a certain abilityCombination) *)
(* TESTS *)
(* ... *)
(* FUNCTION BODY *)
payoffMatrix[abilityKeyCombination_, playerKey_]:=Map[fnPayoff[abilityKeyCombination, #, playerKey]&, emptyMatrix, {n}];
(* COMMENTS
Since it is a function it should actually be called fnPayoffMatrix. In analogy to the data type payoffMatrix I nevertheless dropped the fn ... prefix. *)

(* game *)
(* SIGNATURE           [abilityKeyCombination_] --> game (= payoffMatrixCombination) *)
(* PURPOSE             constructs the payoff-matrix for each player in a certain shape (i.e. for a certain abilityCombination) *)
(* TESTS *)
(* ... *)
(* FUNCTION BODY *)
game[abilityKeyCombination_]:=Map[payoffMatrix[abilityKeyCombination, #]&, Range[n], {1}]
(* COMMENTS
Since it is a function it should actually be called fnGame. In analogy to the data type game I nevertheless dropped the fn ... prefix. *)


(* --------------6---------------- *)
(* feasibilityMatrix construction *)

(* fnFeasibleForPlayerQ *)
(* SIGNATURE           [abilityKey_, actionKey_, playerKey_] --> feasibility *)
(* PURPOSE             test if the combination repaymentKey and abilityKey is possible for a certain player *)
(* TEST *)
(* ... *)
(* FUNCTION BODY *)
fnFeasibleForPlayerQ[abilityKey_, actionKey_, playerKey_]:=Refine[abilitySet[playerKey][[abilityKey]]<=fnGetActionRepayment@(actionSet[playerKey][[actionKey]])];

(* fnDetermineInfeasibleActionKeys *)
(* SIGNATURE           [abilityKey_, playerKey_] --> List of actionKeys *)
(* PURPOSE             get all infeasible actionKeys for a certain abilityKey *)
```

```
(* TESTS *)
(* ... *)
(* FUNCTION BODY *)
fnDetermineInfeasibleActionKeys [ abilityKey_ , playerKey_]:=Module[
{evaluateRobot, infeasibleActionKeys },
(* robot that checks actionKey for feasibility and denotes infeasible Keys *)
evaluateRobot [actionKey_]:= If [ fnFeasibleForPlayerQ [ abilityKey ,actionKey ,playerKey ], True,actionKey];
(* apply robot to all existing actionKeys *)
infeasibleActionKeys =Map[evaluateRobot[#]&,Range[Length[actionSet[playerKey] ]] ];
(* delete True entries for the resulting list *)
infeasibleActionKeys =DeleteCases[ infeasibleActionKeys , True];
(* return resulting list of infeasible actionKeys *)
Return[ infeasibleActionKeys ];
];

(* feasibilityMatrix *)
(* SIGNATURE          [abilityKeyCombination_] --> feasibilityMatrix *)
(* PURPOSE            construct the feasibilityMatrix under a certain abilityCombination , i.e. in a certain shape of the repayment game *)
(* TESTS *)
(* ... *)
(* FUNCTION BODY *)
feasibilityMatrix [abilityKeyCombination_]:=Module[
{ feasibilityMatrix , evaluateRobot , fnCancelRobot },

(* STEP 1: Start out with a booleanMatrix with only True entries *)
feasibilityMatrix =ConstantArray[True,nrOfActionsCombination]; (* Template with True everywhere *)

(* STEP 2: *)
(* robot to cancel positions that are infeasible for a player *)
fnCancelRobot[playerKey_]:=Module[
{ infeasibleActionKeys , infeasiblePositions },
(* Step A: Find infeasible actions for the player *)
infeasibleActionKeys =fnDetermineInfeasibleActionKeys [ abilityKeyCombination [[ playerKey ]], playerKey ];
(* Step B: Transform into infeasible positions *)
infeasiblePositions =Sequence@@ReplacePart[ConstantArray[All,n],playerKey->infeasibleActionKeys];
(* Step C: Replace infeasible positions *)
feasibilityMatrix [[ infeasiblePositions ]]=False;
];

(* STEP 3: apply fnCancelRobot for each playerKey *)
Map[fnCancelRobot[#]&,Range[n]]; (* excecute fnCancelRobot for every player *)
Return[ feasibilityMatrix ]; (* return feasibilityMatrix *)
];
(* COMMENTS
1) Since it is a function it should actually be called fnFeasibilityMatrix . In analogy to the data type feasibilityMatrices I nevertheless dropped the fn ... prefix .
2) By actionwise replacement we save computation effort and time. Instead of calling the fnFeasibleForPlayerQ[] function for every entry in the xxxMatrix i.e. length[ actionSet ]*length[ actionSet ]*... times,
       we need to call it only length[ actionSet ]+length[ actionSet ]+... times *)
```

A.3 Code block: Game solution

A.3.1 Problem statement

Write a software program that can solve discrete games and calculated the Nash-equilibria. More precisely is shall be able to compute pure Nash equilibria in a discrete, non-cooperatively, simultaneously played game. The function must take as input the payoff matrices of the players. In addition it should take a matrix (same dimensions as one payoff matrix) with boolean entries, telling which action combinations are feasible. The Nash-Equilibria shall only be computed based on feasible action combinations. The whole code block here shall be absolutely self-contained and independent from any global variable or functions used in other parts of the code. This makes it reusable outside the world of repayment games. Therefore, a "new" information analysis is to be performed.

A.3.2 Information Analysis

First of all we have a number of players `nrOfPlayers`. Each player has a unique `playerNr`.[116] Notice the number of player is a constant within a game; since we want to write a piece of software that works for arbitrary numbers of players it is in fact a variable in our code.

Each player has of course a set of `actions`, which are addressed by a (within a player's action set unique) `actionNr`. The choice of actions of all players is captured by an `actionCombination`. If we consider a certain action combination, we can have different information about it.

`Payoff`	a number describing the payoff a player receives if a certain action combination is played
`Feasibility`	a boolean value represting the information if a certain action combination can be played, i.e. is feasible
`Best Response`	a boolean value representing the information if a certain action combination is a best response for a certain player
`Nash Equilibrium`	a boolean value telling wether a certain action combination represents a non-cooperative Nash equilibrium
`Solution`	a string with three possible values, one marking an action combination as infeasible, one for marking a combination as Nash Equilibrium, and one for all other cases.

116 Unfortunately, this information is called `playerKey` in the other code blocks. But the term 'Key' is better as it is clearly an ordinal and avoids confusion with cardinal numbers.

`xxxMatrix` represents a compound data structure. This should be seen as being a new 'primitive' data type (although it is of course not a primitve). It simply is a matrix of adequate dimensions, i.e. the dimensions of the game. It does not specify what are the entries are, how they must be interpretated. It is basically a template which can be filled with different information. We will fill a game matrix with concrete atomic data types.

`Payoff-Matrix`	a game matrix of payoffs, telling the payoff a player receives for all action combinations
`Feasibility-Matrix`	a game matrix of boolean values, telling the feasibility of all action combinations
`BestResponse-Matrix`	a game matrix of boolean values, telling if an action combination is a best response for a certain player for all action combinations
`NashEq-Matrix`	a game matrix of boolean values, telling if an action combination is a non-cooperative Nash equilibrium for all action combinations
`Solution-Matrix`	a game matrix of solution strings, telling feasibility and equilibrium properties for each action combination

`Game` is a data type that combines the payoff-Matrices of all players.

A.3.3 Brief description of the algorithm

The difficult task is to transform the Payoff-Matrix of each player into a BestResponse-Matrix (fnMarkBestResponse) whilst excluding infeasible actionCombinations. Once that is done we easily get the Equilibrium actionCombinations by combining these BestResponse-Matrices (fnMarkNashEq). Also the final presentation of the game is easy, we have to check if an actionCombination is feasible, an equilibrium or not and create a xxxMatrix with the respective string at the position of each actionCombination. (fnSolveGame).

To determine best responses of a player we divide the task into four steps.

1. `fnDropFalse` drops infeasible entries in the `payoffMatrix` as these cannot be best responses (technically this is done be replacing such entries by the symbol `Null`).
2. `fnGetReachablePositions` determines the `actionCombinations` a player can reach given his opponents' actions.

3. `fnCalculateMax` calculates the maximal payoff within all reachable entries (ignoring infeasible entries, using global assumptions).
4. `fnMarkEqualToValue` marks the entries that deliver maximal payoff with `True` and all others with `False`, i.e. it basically saves the result of the best response analysis.

The function `fnMarkBestResponses` finally combines all these auxilary functions to transform a whole `payoffMatrix` into a `bestResponseMatrix`.

Notice that information is 'lost' in the course of the algorithm. In a first step concrete `payoffMatrices` are transformed into `booleanMatrices` that just tell if a certain `actionCombination` is a best response for a player. In a second step several `bestResponseMatrices` are condensed into a `nashEqMatrix` that just tells if a certain `actionCombination` is a Nash equilibrium or not.

The main feature of this algorithm is that is operates on whole sets of feasible matrix positions. The reason we do it this way is to save computation time. If on the contrary we checked each position in the payoffMatrix separately whether it is maximal this would mean to calculate the same maximum several times.

Nevertheless, I believe that further improvements should be possible, also with respect to highly optimized internal functions introduced in newer Mathematica versions. Some suggestions are:

- Make `feasibilityMatrix` an optional argument in the functions `fnSolveGame[]` and `fnSolveGameTraceable[]` and set default value to the constant array filled with `True` everywhere (use build-in `Array[]` function).
- This algorithm works quite 'graphically', `fnSolveGame[]` for example returns a whole `xxxMatrix`, although there maybe only one equilibrium. Also the function `fnMarkBestResponse[]` returns a whole `xxxMatrix` although we could aswell only return a list of the best response `actionCombinations` – which would be more memory efficient.
- The function `fnSolveGame[]` could be made more efficient. Currently it first calculates all players best response matrices and then combines them. But positions that are not a best response for the first player cannot be an equilibrium, so we can save the calculation effort of checking whether they are a best response for any other player. (The function `fnSolveGameTraceable[]`, however, needs to calculate all best responses for all players.)

A.3.4 Conventions

Unfortunately commonly used naming conventions cannot be used together with Mathematica symbolic language. Therefore I adopt the following conventions:

Data definition names use upper camel case, starting with a small type to avoid conflicts the Mathematica expressions.

Constants are typeset in ALL-CAPS.

Variables commence with v... and proceed in upper camel case.

Functions commence with fn... and proceed in upper camel case.

Functions returning a single boolean value end with Q (according to the Mathematica convention).

Functions transforming a list of lists into a list with only boolean leaves commence with fnMark... and then indicate how `True` is to be interpreted.

Functions are ordered from auxilary functions to combining functions, as Mathematica has to read the auxilary functions first before other functions can call them. Unfortunately this puts the most important functions last.

A.3.5 Source code

```
(* ========== *)
(* CONSTANTS *)

INF;  (* sympol for impossible action-combinations, without value *)
EQU;  (* symbol for Nash equilibrium action-combinations, without value *)
ELS;  (* symbol for Not Nash equilibrium action-combinations, without value *)

(* ================= *)
(* DATA DEFINITIONS *)

(* nrOfPlayers *)
(* DEFINITION          is positive Integer *)
(* INTERPRETATION      the total number of players *)
(* EXAMPLES            n=2; n=5; *)
(* TEMPLATE            fnForN[n_]:= ... n ; *)
(* RULES USED          atomic non-distinct: positive Integer *)

(* playerNr *)
(* DEFINITION          is Integer [1,nrOfPlayers] *)
(* INTERPRETATION      the number of a certain player *)
(* EXAMPLES            playerNrMax=1; playerNrMoritz=2; *)
(* TEMPLATE            fnForPlayerNr[playerNr_]:= ... playerNr ; *)
(* RULES USED          atomic non-distinct: Integer [1,nrOfPlayers] *)

(* actionNr *)
(* DEFINITION          is Integer [1,nrOfPlayers] *)
(* INTERPRETATION      the number of an action a specified player can play *)
(* EXAMPLES            actionRepay=1; actionDefault=2 *)
(* TEMPLATE            fnForAction[ action_]:= ... action ; *)
(* RULES USED          atomic non-distinct: Integer [1,nrOfPlayers] *)

(* actionCombination *)
(* DEFINITION          is List of positive Integers *)
(* INTERPRETATION      an action combination in the game, the first entry denotes the actionNr of the first player etc. *)
(* EXAMPLES            actionCombinationA={1,1,1}; actionCombinationB={1,1,2} *)
(* TEMPLATE            fnForActionCombination[actionCombination_]:= ... actionCombination [[...]] ; *)
(* RULES USED          compound: nrOfPlayers fields *)

(* payoff *)
(* DEFINITION          is Real (or a Mathematica expression representing a Real number) *)
(* INTERPRETATION      the payoff a player receives if a certain action combination is played *)
(* EXAMPLES            payoffCombination111Player1=7; payoffCombination111Player2=-2R-p *)
(* TEMPLATE            fnForPayoff[ payoff_]:= ... payoff ; *)
(* RULES USED          atomic non-distinct: Real number *)

(* feasibility *)
(* DEFINITION          is Boolean *)
(* INTERPRETATION      true if action combination is feasible , false otherwise *)
(* EXAMPLES            feasibilityCombination111 =True; pfeasibilityCombination112 =False *)
(* TEMPLATE            fnForFeasibility [ feasibility_ ]:= ... feasbility ; *)
(* RULES USED          atomic non-distinct: Boolean *)
```

```
(* bestResponse *)
(* DEFINITION         is Boolean *)
(* INTERPRETATION     true if action combination is a best response for a player (to be specified), false otherwise *)
(* EXAMPLES           bestResponseCombination111Player1=True; bestResponseCombination111Player2=False *)
(* TEMPLATE           fnForBestResponse[ bestresponse_]:= ... bestResponse ; *)
(* RULES USED         atomic non-distinct: Boolean *)

(* nashEq *)
(* DEFINITION         is Boolean *)
(* INTERPRETATION     true if action combination is a NashEquilibrium, false otherwise *)
(* EXAMPLES           nashEqCombination111=True; nashEqCombination112=False *)
(* TEMPLATE           fnForNashEq[nashEq_]:= ... nashEq ; *)
(* RULES USED         atomic non-distinct: Boolean *)

(* solution *)
(* DEFINITION         is one of INF,EQU,ELS *)
(* INTERPRETATION     the result of the evaluation of a certain action combination: infeasible , equilibrium , else *)
(* EXAMPLES           solutionCombination111=INF; nashequilibriumCombination112=EQU *)
(* TEMPLATE           fnForSolution[ solution_]:= Which[solution==INF, ... , solution==EQU, ... , solution==ELS, ...] ; *)
(* RULES USED         one of: 3 cases; atomic distinct : INF; atomic distinct : EQU; atomic distinct : ELS *)

(* xxxMatrix *)
(* DEFINITION         is List of Lists of Lists ... of the dimensions of the game with content xxx (of arbitrary type) *)
(* INTERPRETATION     placeholder for entries of the data types defined above *)
(* EXAMPLES           payoffMatrix=payoffsOfPlayer1; feasibilityMatrix =feasibleActionCombinations *)
(* TEMPLATE           fnForGameMatrix[xxxMatrix_]:= ... xxxMatrix [[...]](* entry*) ; *)
(* RULES USED         compound: many fields (according to the dimensions of the game) *)

(* game *)
(* DEFINITION         is List of payoffMatrices *)
(* INTERPRETATION     the first payoffMatrix describes the payoffs of the first player ... etc. *)
(* EXAMPLES           game1 = {{{2, 3}, {5, 1}}, {{6, 2.1}, {13, 2}}}; *)
(* TEMPLATE           fnForGame[game_]:=... game [[...]] (*payoffMatrix*); *)
(* RULES USED         compound: nrOfPlayers fields ; xxxMatrix; payoff; *)


(* ============ *)
(* FUNCTIONS *)


(* fnDropFalse *)
(* SIGNATURE          [payoffMatrix_, feasibilityMatrix_ ] --> payoffMatrix *)
(* PURPOSE            replace entries of infeasible action combinations in payoffMatrix by Null *)
(* TESTS *)
fnDropFalse[{{1, 2, 3}, {4, 5, 6}}, {{True, True, False}, {True, False, True}}] === {{1, 2, Null}, {4, Null, 6}};
fnDropFalse[{{-5, 2}, {7.1, Pi}}, {{True, True}, {True, True}}] === {{-5, 2}, {7.1, Pi}};
fnDropFalse[{{-5, 2}, {7.1, Pi}}, {{False, False}, {False, False}}] === {{Null, Null}, {Null, Null}};
fnDropFalse[{{{1, 2}, {3, 4}}, {{5, 6}, {7, 8}}}, {{{True, False}, {True, True}}, {{True, True}, {True, True}}}] === {{{1, Null}, {3, 4}}, {{5, 6}, {7, 8}}};
(* FUNCTION BODY *)
fnDropFalse[xxxMatrix_, feasibilityMatrix_ ]:=Module[
{fnReplaceRobot,vReplacedMatrix},
fnReplaceRobot[value_, position_]:= If[ feasibilityMatrix [[Sequence@@position]]==False,Null,value]; (* function that decides upon replacement *)
```

```
vReplacedMatrix=MapIndexed[ fnReplaceRobot[#1,#2]&,xxxMatrix,{ArrayDepth[feasibilityMatrix]}];
(* fnReplaceRobot[] applied to entries of payoffMatrix *)
Return[vReplacedMatrix]; (* return resulting matrix *) ];


(* fnGetReachablePositions *)
(* SIGNATURE          [actionCombination_, playerNr_] --> List *)
(* PURPOSE            determine which positions a player can reach under given actions of his opponents *)
(* TESTS *)
fnGetReachablePositions [{1,1},1]==={All,1};
fnGetReachablePositions [{1,2,5,3,2},4]==={1,2,5,All,2};
(* FUNCTION BODY *)
fnGetReachablePositions [actionCombination_, playerNr_]:=ReplacePart[actionCombination,playerNr->All];


(* fnCalculateMax *)
(* SIGNATURE          [ listOfPayoffs_ ] --> Real (or Mathematica symbol representing a Real number) *)
(* PURPOSE            determine the maximum in a list of (maybe symbolic) payoffs *)
(* TESTS *)
fnCalculateMax[{2,Pi,5,12.3}]===12.3;
fnCalculateMax[{Null,1,5,1,Null}]===5;
Assuming[e>R,fnCalculateMax[{R,e,Null}]===e];
(* FUNCTION BODY *)
fnCalculateMax[ listOfPayoffs_ ]:=Module[
{vList,vMax},
vList= listOfPayoffs ; (* Initialization *)
vList=DeleteCases[vList,Null]; (* Delete entries that are Null *)
vMax=Max[vList]; (* Calculate Maximum *)
vMax=Refine[vMax]; (* Refine the result using global assumptions *)
Return[vMax];];


(* fnMarkEqualToValue *)
(* SIGNATURE          [ listOfValues_, value_] --> [listOfBoolean] *)
(* PURPOSE            transform a list of values into a list of boolean entries , with true at positions that coincide with value_ *)
(* TESTS *)
fnMarkEqualToValue[{3,4,5},4]==={False,True,False};
fnMarkEqualToValue[{1,2,1,1,3},1]==={True,False,True,True,False};
fnMarkEqualToValue[{1,Null,1,1,3},1]==={True,False,True,True,False};
(* FUNCTION BODY *)
fnMarkEqualToValue[listOfValues_, value_]:=Module[
{fnReplaceRobot,vReplacedList},
(* function that decides upon replacement *)
fnReplaceRobot[entry_]:=Which[
entry===Null,            False,
Refine[entry==value],    True,
True,                    False];
(* fnReplaceRobot[] applied to the entries of listOfValues *)
vReplacedList=Map[fnReplaceRobot[#]&,listOfValues];
(* return resulting list *)
Return[vReplacedList]; ];
```

```
(* fnMarkBestResponse *)
(* SIGNATURE          [ payoffMatrix_, feasibilityMatrix_ , playerNr_] --> bestResponseMatrix *)
(* PURPOSE            transform a payoffMatrix into booleanMatrix with True if a position is a best response *)
(* TESTS *)
fnMarkBestResponse[{ {2, 3}, {4, 2} }, { {True, False}, {True, True} }, 1] === { {False, False}, {True, True} };
fnMarkBestResponse[{ {1, 2, 3}, {4, 5, 6} }, { {True, True, False}, {True, False, True} }, 2] === { {False, True, False}, {False, False, True} };
fnMarkBestResponse[{ { {1, 2}, {3, 4} }, { {5, 6}, {7, 8} } }, { { {True, True}, {True, True} }, { {True, False}, {True, True} } }, 1] ===
{ { {False, True}, {False, False} }, { {True, False}, {True, True} } };
(* FUNCTION BODY *)
fnMarkBestResponse[payoffMatrix_, feasibilityMatrix_ , playerNr_] :=Module[
{ vPayoffMatrix,fnReplaceRobot,fnReplaceRobotSmart,vResult },
(* Preparation of the payoff Matrix *)
vPayoffMatrix=fnDropFalse[ payoffMatrix , feasibilityMatrix ]; (* Drop entries of infeasible action combinations *)
(* Robot function to mark best responses in a certain situation *)
fnReplaceRobot[actionCombination_]:=Module[
{ vSelectorSequence, vReachableEntries ,vMax },
vSelectorSequence=Sequence@@fnGetReachablePositions[actionCombination,playerNr]; (* get reachable positions *)
vReachableEntries=vPayoffMatrix[[ vSelectorSequence ]]; (* get list of reachable entries *)
vMax=fnCalculateMax[vReachableEntries]; (* calculate maximum *)
vReachableEntries=fnMarkEqualToValue[vReachableEntries,vMax]; (* replace list of reachable entries by boolean values with True for maximum *)
vPayoffMatrix [[ vSelectorSequence ]]= vReachableEntries ; (* substitute reachable entries boolean version into vPayoffMatrix *)];
(* Making the robot smart by applying it only to new situations for a player *)
fnReplaceRobotSmart[actionCombination_]:= If [actionCombination [[ playerNr ]]==1, fnReplaceRobot[actionCombination ]];
(* Applying the robot to the payoffMatrix *)
MapIndexed[ fnReplaceRobotSmart[#2]&,vPayoffMatrix, { ArrayDepth[vPayoffMatrix] } ];
(* Return vPayoffMatrix *)
Return[vPayoffMatrix ];];


(* fnMarkNashEq *)
(* SIGNATURE          [game_ feasibilityMatrix_ ] --> nashEqMatrix *)
(* PURPOSE            transform a game into a booleanMatrix with True if a position is a Nash equilibrium *)
(* TESTS *)
fnMarkNashEq[ { { {2, 3}, {4, 2} }, { {5, 7}, {2, 3} } }, { {True, False}, {True, True} } ] === { {False, False}, {False, True} };
(* FUNCTION BODY *)
fnMarkNashEq[game_, feasibilityMatrix_ ]:=Module[
{ vNrOfPlayers,vBestResponseMatrices,vNashEqMatrix },
(* Determine number of players in the game *)
vNrOfPlayers=Length[game];
(* produce vBestResponseMatrices by applying fnMarkBestResponse to all payoffMatrices of the game *)
vBestResponseMatrices=Map[fnMarkBestResponse[game[[#]],feasibilityMatrix ,#]&,Range[vNrOfPlayers]];
(* Combine bestResponses to NashEquilibria *)
vNashEqMatrix=MapThread[And,vBestResponseMatrices,vNrOfPlayers];
(* Return vNashEqMatrix *)
Return[vNashEqMatrix];];


(* fnSolveGame *)
(* SIGNATURE          [game_ feasibilityMatrix_ ] --> solutionMatrix *)
(* PURPOSE            mark the actionCombinations in a xxxMatrix with one of INF,EQU,ELS *)
(* TESTS *)
fnSolveGame[ { { {2, 3}, {4, 2} }, { {5, 7}, {2, 3} } }, { {True, False}, {True, True} } ] === { {ELS, INF}, {ELS, EQU} };
(* FUNCTION BODY *)
```

```
fnSolveGame[game_, feasibilityMatrix_ ]:=Module[
{vNashEqMatrix,fnRobot,vSolutionMatrix},
(* Find Nash Equilibria *)
vNashEqMatrix=fnMarkNashEq[game,feasibilityMatrix];
(* Robot that treats each position accordingly *)
fnRobot[actionCombination_]:=Which[
! feasibilityMatrix [[Sequence@@actionCombination]], INF,
vNashEqMatrix[[Sequence@@actionCombination]],                EQU,
True,                                                                        ELS];
(* Apply fnRobot to each actionCombination *)
vSolutionMatrix=MapIndexed[fnRobot[#2]&,game[[1]],{ArrayDepth[game[[1]]]}];
(* Return vSolutionMatrix *)
Return[vSolutionMatrix ];];


(* fnSolveGameTraceable *)
(* SIGNATURE        [game_, feasibilityMatrix_ ] --> xxxMatrix *)
(* PURPOSE          produce a solution that is traceable by humans, entries xxx are payoffCombinations with best responses underlined *)
(* TESTS *)
(* FUNCTION BODY *)
fnSolveGameTraceable[game_, feasibilityMatrix_ ] :=Module[
{vNrOfPlayers,vBestResponseMatrices,fnUnderlineRobot,fnUnderlineRobots,vUnderlinedGame,vResult},
(* Determine number of players in the game *)
vNrOfPlayers=Length[game];
(* produce vBestResponseMatrices by applying fnMarkBestResponse to all payoffMatrices of the game *)
vBestResponseMatrices=Map[fnMarkBestResponse[game[[#]],feasibilityMatrix,#]&,Range[vNrOfPlayers]];
(* Robot that underlines best responses in the payoffMatrices *)
fnUnderlineRobot[booleanMatrix_, value_, position_ ]:= If[booleanMatrix[[Sequence@@position]],Style[value,Underlined], Style[value]];
(* clone the robot apply them to a whole payoffMatrix *)
fnUnderlineRobots[ payoffMatrix_,bestResponseMatrix_]:=MapIndexed[fnUnderlineRobot[bestResponseMatrix,#1,#2]&,payoffMatrix,{ArrayDepth[payoffMatrix]}];
(* apply the cloned robots to the whole game *)
vUnderlinedGame=Map[fnUnderlineRobots[game[[#]],vBestResponseMatrices[[#]]]&,Range[vNrOfPlayers]];
(* combine the payoffMatrices to one Matrix with entries payoffCombinations *)
vResult=MapThread["",vUnderlinedGame,ArrayDepth[vUnderlinedGame[[1]] ]];
(* Drop infeasible actionCombinations in the Matrix *)
vResult=fnDropFalse[vResult, feasibilityMatrix ]/. Null->INF;
(* Return result *)
Return[vResult];
];
```

A.4 Code block: Game analysis

A.4.1 Problem statement

These functions shall take the `solutionMatrix` of a certain `abilityCombination` and calculate information about the equilibria found in the `solutionMatrix`. To perform these calculation they will have to refer to the global variables described on page p. 203 and combine this information with the information from the `solutionMatrix`. We want to calculate:

1. Number of equilibria
2. To build the map of equilibria: equilibrium actionNameCombinations
3. To build the map of outcomes: equilibrium payoffCombinations
4. To build the map of repayment: equilibrium sum of repayments
5. To built the map of punishment load: equilibrium total punishment load (punishment and sanctions of all borrowers)

In addition we want to have a method to display what the punishment strategy resulting from our `punishmentFunction` is. So there shall be a function that returns a `xxxTos` with entries the punishment applied in each combination. Or more general: a function that returns a `xxxTos` with entries the `payoffLayerCombinations` for a `payoffLayer` of our choice.

The methods should not use any data types other than defined in the game construction block.

A.4.2 Brief description of the algorithm

The methods defined in this code block are rather self-explanatory.

Maybe the code could be made more elegant by using more abstract 2nd order extractor functions, because the functions here are so similar with lots of redundant code. For example, write a function that extracts equilibrium Combinations and then functions that turn combinations to repaymentCombination, or combination to punishmentCombination, etc.

A.4.3 Source code

```
(* ========== *)
(* FUNCTIONS *)

(* fnGetEqActions *)
(* SIGNATURE        [ solutionMatrix_ ] --> actionNameCombinationsList *)
(* PURPOSE          get the equilibrium actionCombinations of a game, to build the map of equilibria *)
(* TESTS *)
(* ... depend on global variable actionSet[i] ... *)
(* fnGetEqActions[{{EQU,ELS,ELS,ELS,INF,INF},{ELS,ELS,ELS,ELS,INF,INF},{ELS,ELS,ELS,ELS,INF,INF},{ELS,ELS,ELS,EQU,INF,INF},{INF,INF,INF,INF,INF,INF},{INF,INF,INF,INF,INF,INF}}]==="0R-n|0
        R-n\n1R-y|1R-y" *)
(* FUNCTION BODY *)
fnGetEqActions[ solutionMatrix_ ]:=Module[
{ result },
(* Extraction of equilibrium actionKeyCombinations *)
result =Position[ solutionMatrix ,EQU];
(* Get from actionKeys to actions *)
result =Map[fnKeyToActionCombination,result];
(* Get form actions to actionNames *)
result =Map[fnGetActionName,result, {2}];
(* Formatting *)
result =fnFormatCombinationsList[ result ];
(* return result *)
Return[ result ];
];


(* fnGetLenderPayoffPerPlayer *)
(* SIGNATURE        [abilityKeyCombination_, actionKeyCombination_,playerKey_] --> lenderPayoff *)
(* PURPOSE          calculate a lenders payoff from one borrower in a particular situation *)
(* TESTS *)
(* depend on global variables punishmentFunction, sanctionFunction, actionSets *)
(* FUNCTION BODY *)
fnLenderPayoffPerPlayer [abilityKeyCombination_, actionKeyCombination_,playerKey_]:=Module[
{abilityCombination, actionCombination, payoff },
abilityCombination =fnKeyToAbilityCombination @abilityKeyCombination; (* get from abilityKeys to abilities , since punishmentFunction and sanctionFunction need these *)
actionCombination=fnKeyToActionCombination@actionKeyCombination; (* get from actionKeys to actions , since punishmentFunction and sanctionFunction need these *)
payoff=-fnGetActionRepayment@(actionCombination[[playerKey]])
-rewardFunction[abilityCombination ,actionCombination ,playerKey ]; (* calculate payoff *)
Return[payoff]; (* return calculated payoff *)
];


(* fnGetLenderPayoffTotal *)
(* SIGNATURE        [abilityKeyCombination_, actionKeyCombination_] --> lenderPayoff *)
(* PURPOSE          calculate a lenders total payoff in a particular situation *)
(* TESTS *)
(* depend on global variables punishmentFunction, sanctionFunction, actionSets *)
(* FUNCTION BODY *)
fnLenderPayoffTotal [abilityKeyCombination_, actionKeyCombination_]:=Plus@@Map[fnLenderPayoffPerPlayer[abilityKeyCombination,actionKeyCombination,#]&,Range[n], {1}];

(* fnGetEqLenderPayoffTotal *)
```

```
(* SIGNATURE        [abilityKeyCombination_, solutionMatrix_] --> lenderPayoffList *)
(* PURPOSE          calculate lenderPayoff in equilibrium, to build the map of lenderPayoff *)
(* TESTS *)
(* depends on global variables *)
(* FUNCTION BODY *)
fnGetEqLenderPayoffTotal[abilityKeyCombination_, solutionMatrix_]:=Module[
{eqActionKeyCombinationsList, lenderPayoffTotal},
(* Extraction of equilibrium actionKeyCombinations *)
eqActionKeyCombinationsList=Position[solutionMatrix,EQU];
(* get lenderPayoffTotal from all equilibrium actionKeyCombinations *)
lenderPayoffTotal=Map[fnLenderPayoffTotal[abilityKeyCombination,#]&,eqActionKeyCombinationsList,{1}];
(* return result *)
Return[lenderPayoffTotal];
];


(* fnGetEqPayoffs *)
(* SIGNATURE        [abilityKeyCombination_, solutionMatrix_] --> payoffCombinationsList *)
(* PURPOSE          get the equilibrium payoff for each borrower, to calculate expected return (meta-game relevant) *)
(* TESTS *)
(* depends on global variables *)
(* FUNCTION BODY *)
fnGetEqPayoffs[abilityKeyCombination_, solutionMatrix_]:=Module[
{eqActionKeyCombinationsList,robotGetPayoffCombination,payoffCombinationsList},
(* Extraction of equilibrium actionKeyCombinations *)
eqActionKeyCombinationsList=Position[solutionMatrix,EQU];
(* robot to get payoffCombination from actionKeyCombination *)
robotGetPayoffCombination[actionKeyCombination_]:=Map[fnPayoff[abilityKeyCombination,actionKeyCombination,#]&,Range[n],{1}];
(* apply robot *)
payoffCombinationsList=Map[robotGetPayoffCombination[#]&,eqActionKeyCombinationsList,{1}];
(* return result *)
Return[payoffCombinationsList];
];


(* fnGetLossSum *)
(* SIGNATURE        [abilityKeyCombination_, actionKeyCombination_] --> deadweightList *)
(* PURPOSE          get the deadweight-loss in a certain situation *)
(* TESTS *)
(* FUNCTION BODY *)
fnGetLossSum[abilityKeyCombination_,actionKeyCombination_]:=Module[
{abilityCombination, actionCombination, loss, lossCombination, lossSum},
abilityCombination=fnKeyToAbilityCombination@abilityKeyCombination; (* get from abilityKeys to abilities, since punishmentFunction and sanctionFunction need these *)
actionCombination=fnKeyToActionCombination@actionKeyCombination; (* get from actionKeys to actions, since punishmentFunction and sanctionFunction need these *)

loss[playerKey_]:= punishmentFunction[abilityCombination,actionCombination,playerKey]
+ sanctionFunction[abilityCombination,actionCombination,playerKey]; (* calculate loss of a player *)
lossCombination=Map[loss[#]&,Range[n],1]; (* calculate loss for all players *)
lossSum=Plus@@lossCombination; (* calculate the total loss, i.e. the sum of the loss of each player *)

Return[lossSum]; (* return calculated payoff *)
];
```

```
(* fnGetEqLoss *)
(* SIGNATURE         [ abilityKeyCombination_, solutionMatrix_ ] --> lenderPayoffList *)
(* PURPOSE           calculate lenderPayoff in equilibrium, to build the map of lenderPayoff *)
(* TESTS *)
(* depends on global variables *)
(* FUNCTION BODY *)
fnGetEqLoss[abilityKeyCombination_, solutionMatrix_ ]:=Module[
{eqActionKeyCombinationsList,loss},
(* Extraction of equilibrium actionKeyCombinations *)
eqActionKeyCombinationsList=Position[ solutionMatrix ,EQU];
(* get lenderPayoffTotal from all equilibrium actionKeyCombinations *)
loss=Map[fnGetLossSum[abilityKeyCombination,#]&,eqActionKeyCombinationsList, {1}];
(* return result *)
Return[loss];
];


(* fnGetPayoffLayerCombinationsTos *)
(* SIGNATURE         xxxFunction --> xxxTos *)
(* PURPOSE           builds payoffLayersTos from the input layerFunction, used to illustrate the punishment strategy defined by punishmentFunction *)
(* TESTS *)
(* ... *)
(* FUNCTION BODY *)
fnGetPayoffLayerCombinationsTos[ layerFunction_ ]:=Module[
{layerFunctionWithKeys,layerCombination, layerMatrix, layerMatrixFeasible , layerTos},
(* transform actionKeyCombination to actionCombination as layerFunction needs it that way *)
layerFunctionWithKeys[ abilityKeyCombination_,actionKeyCombination_,playerKey_]:= layerFunction [fnKeyToAbilityCombination @abilityKeyCombination,fnKeyToActionCombination@actionKeyCombination,playerKey];
(* Build layerCombinations from individual layer entries per player *)
layerCombination[ abilityKeyCombination_, actionKeyCombination_]:=Map[layerFunctionWithKeys[abilityKeyCombination,actionKeyCombination,#]&,Range[n],1] ;
(* Build layerMatrices from layerCombinations *)
layerMatrix [ abilityKeyCombination_]:=Map[layerCombination[abilityKeyCombination,#]&,emptyMatrix, {n}];
(* Drop infeasible entries from layerMatrix *)
layerMatrixFeasible [ abilityKeyCombination_]:=fnDropFalse[ layerMatrix [ abilityKeyCombination ], feasibilityMatrix [ abilityKeyCombination ]];
(* Build layerTos from layerMatrices *)
layerTos=Map[layerMatrixFeasible[#]&,emptyTos, {n}];
Return[layerTos];
];
```

A.5 Code block: Game presentation

A.5.1 Problem statement

Write a method that take an `xxxMatrix` of arbitrary dimension and present it nicely and easy to read. In particular the output should have headings, such that each entry can be easily attributed to the respective `actionCombination`. Likewise, write a method to display a tabular of shape, i.e. one `xxxMatrix` for every `abilityCombination`.

Furthermore, include any helpful functions to generate a nice graphical output in this code block here.

The functions here shall only refer to the global variables `n`, `actionSet[i]`, `abilitySet[i]`, and use only methods from the game construction block. This way the code block remains independent of the blocks game solution and game analysis.

A.5.2 Brief description of the algorithm

The algorithm to display a `xxxMatrix` or a `xxxTos` proceeds in two steps. Step1: Construct Headings. This part collects the necessary data for the headings and formats it. Step2: Display `xxxMatrix/xxxTos`. This part actually formats and displays a given `xxxMatrix/xxxTos`.

A.5.3 Source code

```
(* ========== *)
(* CONSTANTS *)

cCombinationEntrySeparator = "|"; (* Separates the entries of xxxCombination *)
cCombinationSeparator = "\n"; (* Separates the entries of xxxCombinationsList *)

cPlayerActionSeparator =":"; (* Separates playerKey and action in the headings of an xxxMatrix *)
cPlayerAbilitySeparator = ":"; (* Separates playerKey and ability in the headings of an xxxTos *)

cHeadlineStyle={Bold,Underlined,Larger};
cActionHeadingsStyle = {Smaller, Bold}; (* Style of actionHeadings *)
cAbilityHeadingsStyle = {Bold}; (* Style of abilityHeadings *)

cMatrixStyle={Bold,Smaller}; (* Style of xxxMatrix entries *)
cTosStyle={ Italic }; (* Style of xxxTos entries *)

cInfString = "*"; (* String to mark infeasible actionCombinations in an xxxMatrix *)
cEquString = "\[SixPointedStar]";(* String to mark equilibrium actionCombinations in an xxxMatrix *)
cElsString = "-";(* String to mark all other actionCombinations in an xxxMatrix *)


(* ======================== *)
(* ADDITIONAL DATA TYPES *)

(* actionHeading *)
(* DEFINITION          list of {playerKey,actionName} *)
(* INTERPRETATION      the heading for a column/row in an xxxMatrix; denoting the player who "owns" this dimension and the column/rows actionName *)
(* EXAMPLES            actionHeading1={1,"repay"}, actionHeading2={5,"2R-n"} *)
(* TEMPLATE            fnForActionHeading[actionHeading_]:= ... actionHeading[[1]], actionHeading[[2]]; *)
(* RULES USED          compound: 2 fields ; references : playerKey, actionName *)

(* abilityHeading *)
(* DEFINITION          list of {playerKey, ability } *)
(* INTERPRETATION      the heading for a column/row in an xxxTos,; denoting the player who "owns" this dimension and the column/rows ability *)
(* EXAMPLES            abilityHeading1={1,2R}, abilityHeading2={5,0} *)
(* TEMPLATE            fnForAbilityHeading[ abilityHeading_]:= ... abilityHeading [[1]], abilityHeading [[2]]; *)
(* RULES USED          compound: 2 fields ; references : playerKey, ability *)

(* xxxCombinationsList *)
(* DEFINITION          a list of xxxCombinations *)
(* INTERPRETATION      a list of xxxCombinations *)
(* EXAMPLES            CombinationsList1={{{"A:pl1","A:pl2","A:pl3"},{"B:pl1","B:pl2","B:pl3"}} *)
(* TEMPLATE            fnForxxxCombinationsList[xxxCombinationsList_]:= ... xxxCombinationsList[[item,playerKey]]; *)
(* RULES USED          compound: arbitrary many fields ; references : xxxCombinations *)


(* ========== *)
(* FUNCTIONS *)

(* ------------------------- *)
(* Format section headings *)
```

```
headline[ text_ ]:= Style["\n"<>text,cHeadlineStyle]; (* a simple tag to print healines *)

(* ---------------------------- *)
(* Construct tabular headings *)

(* fnGetActionNameSet *)
(* SIGNATURE          actionSet --> actionNameSet *)
(* PURPOSE            construct action headings for presentation *)
(* TESTS *)
fnGetActionNameSet[{ {"opt1", 0, "n"}, {"opt2", 0, "y"}, {"opt3", -R, "y"}, {"opt4", -2 R, "y"}}] === {"opt1", "opt2", "opt3", "opt4"};
(* FUNCTION BODY *)
fnGetActionNameSet[actionSet_] := fnGetActionName /@ actionSet;

(* fnGetActionHeadings *)
(* SIGNATURE          [actionNameSet_, playerKey_] --> actionHeadingSet *)
(* PURPOSE            construct actionNameHeadingSet, i.e. actionNameSet with player denoted; this function determines how they are formated *)
(* TESTS *)
(* fnGetActionHeadings[fnGetActionNameSet[ actionSet [1]],1]; *)
(* FUNCTION BODY *)
fnGetActionHeadings[actionNameSet_, playerKey_] := Map[(ToString["P"] <> ToString[playerKey] <> cPlayerActionSeparator <> ToString[#]) &, actionNameSet, {1}];

(* fnGetAbilityHeadings *)
(* SIGNATURE          [ abilitySet_ , playerKey_] --> abilityHeadingSet *)
(* PURPOSE            construct abilityHeadingSet for a player; this function determines how they are formated *)
(* TESTS *)
(* fnGetAbilityHeadings[ abilitySet [1],1]; *)
(* FUNCTION BODY *)
fnGetAbilityHeadings[ abilitySet_ , playerKey_] := Map[(ToString["A"] <> ToString[playerKey] <> cPlayerAbilitySeparator <> ToString[#]) &, abilitySet, {1}];

(* actionHeadingsCombination *)
(* SIGNATURE          [ ] --> actionHeadingsCombinarion *)
(* PURPOSE            combine the actionNameSets of each player *)
(* TESTS *)
(* actionHeadingsCombination; *)
(* FUNCTION BODY *)
actionHeadingsCombination := Module[
{vHeadings},
(* Step 1: Construct actionHeadings for all players *)
vHeadings = Map[fnGetActionHeadings[fnGetActionNameSet[actionSet[#]], #] &, Range[n], {1}];
(* Step 2: Format Headings *)
vHeadings = Map[Style[#, cActionHeadingsStyle] &, vHeadings, {2}];
(* Return result *)
Return[vHeadings]; ];

(* abilityHeadingsCombination *)
(* SIGNATURE          [ ] --> abilityHeadingsCombinarion *)
(* PURPOSE            combine the abilityName of each player *)
(* TESTS *)
(* abilityHeadingsCombination; *)
(* FUNCTION BODY *)
abilityHeadingsCombination := Module[
{vHeadings},
```

```
(* Step 1: Construct abilityHeadings for all players *)
vHeadings = Map[fnGetAbilityHeadings[ abilitySet [#], #] &, Range[n], {1}];
(* Step 2: Format Headings *)
vHeadings = Map[Style[#, cAbilityHeadingsStyle ] &, vHeadings, {2}];
(* Return result *)
Return[vHeadings]; ];


(* ----------------------------------- *)
(* Display and format xxxMatrix, xxxTos *)

(* fnDisplayMatrix *)
(* SIGNATURE         xxxMatrix --> Output *)
(* PURPOSE           display xxxMatrices in nice Mathematica Output Form *)
(* TESTS *)
(* ... *)
(* FUNCTION BODY *)
fnDisplayMatrix [xxxMatrix_] := Module[
{xxxMatrixStyled,output},
(* Format entries *)
xxxMatrixStyled=Map[Style[#,cMatrixStyle]&,xxxMatrix,{n}];
(* Plots the tabularOfShapes in TableForm with the following options ... *)
output = TableForm[xxxMatrixStyled,
TableAlignments -> Left, (* how to align entries in each dimension *)
TableDirections -> If[EvenQ[n], Column, Row], (* whether to arrange dimensions as rows or columns *)
TableHeadings -> actionHeadingsCombination, (* how to label table entries *)
TableSpacing -> 1 ]; (* how many spaces to put between entries in each dimension *)
(* Return result *)
Return[output]; ];

(* fnDisplayTos *)
(* SIGNATURE         xxxTos --> Output *)
(* PURPOSE           present all shapes at once *)
(* TESTS *)
(* ... *)
(* FUNCTION BODY *)
fnDisplayTos [xxxTos_] := Module[
{xxxTosStyled,output},
(* Format entries *)
xxxTosStyled=Map[Framed[Style[#,cTosStyle]]&,xxxTos,{n}];
(* Plots the tabularOfShapes in TableForm with the following options ... *)
output = TableForm[xxxTosStyled,
TableAlignments -> Center, (* how to align entries in each dimension *)
TableDirections -> If[EvenQ[n], Column, Row], (* whether to arrange dimensions as rows or columns *)
TableHeadings -> abilityHeadingsCombination, (* how to label table entries *)
TableSpacing -> 3];(* how many spaces to put between entries in each dimension *)
(* Return result *)
Return[output]; ];


(* ---------------------------------------------------------------------- *)
(* Format xxxCombinations, xxxCombinationsLists, solutionMatrix *)
```

```

(* fnFormatCombination *)
(* SIGNATURE          xxxCombination --> String *)
(* PURPOSE            format a xxxCombination using the constant cCombinationEntrySeparator *)
(* TESTS *)
fnFormatCombination[{"A:pl1", "A:pl2", "A:pl3"}] === "A:pl1|A:pl2|A:pl3";
(* FUNCTION BODY *)
fnFormatCombination[xxxCombination_] := StringReplace[ToString[xxxCombination], {
"," -> cCombinationEntrySeparator, (* separate entries *)
"{" -> "", (* delete opening parenthese *)
"}" -> "", (* delete closing parenthese *)
"␣" -> ""}]; (* delete remanin white spaces *)

(* fnFormatCombinationsList *)
(* SIGNATURE          xxxCombinationsList --> String *)
(* PURPOSE            format a list of xxxCombinations using the constants cCombinationEntrySeparator and
cCombinationSeparator *)
(* TESTS *)
fnFormatCombinationsList[{{"A:pl1","A:pl2", "A:pl3"}, {"B:pl1", "B:pl2", "B:pl3"}, {"C:pl1", "C:pl2", "C:pl3"}}] === "A:pl1|A:pl2|A:pl3\nB:pl1|B:pl2|B:pl3\nC:pl1|C:pl2|C:pl3";
(* FUNCTION BODY *)
fnFormatCombinationsList[xxxCombinationsList_] := Module[
{vFormattedCombiList},
(* Step 1: format xxxCombinations using the function fnFormatCombination *)
vFormattedCombiList = Map[fnFormatCombination[#] &, xxxCombinationsList, {1}];
(* Step 2: format the list of formated xxxCombinations usind the constant cCombinationSeparator *)
vFormattedCombiList = ToString[vFormattedCombiList]; (* make everything a string *)
vFormattedCombiList = StringReplace[vFormattedCombiList, {
"," -> cCombinationSeparator, (* separate entries *)
"{" -> "", (* delete opening parenthese *)
"}" -> "", (* delete closing parenthese *)
"␣" -> ""}]; (* delete remaining white spaces *)
(* Return result *)
Return[vFormattedCombiList]; ];

(* fnFormatSolutionMatrix *)
(* SIGNATURE          solutionMatrix --> StringMatrix *)
(* PURPOSE            replace the symbols in a solution Matrix by graphically nice strings using the constants cInfString , cEquString, cElsString *)
(* TESTS *)
fnFormatSolutionMatrix[{{EQU, ELS, INF}, {ELS, ELS, INF}, {INF, INF, INF}}] == {{cEquString, cElsString, cInfString}, {cElsString, cElsString, cInfString }, {cInfString , cInfString , cInfString }};
(* FUNCTION BODY *)
fnFormatSolutionMatrix[ solutionMatrix_ ] := solutionMatrix /. {INF -> cInfString, EQU -> cEquString, ELS -> cElsString};
```

A.6 Code block: Game definition

Now, that all necessary methods are defined and loaded into the Mathematica kernel, we can define the game and use them so solve it.

This piece of code serves as the actual 'user interface'. Here the game definition is written down and the output of the results is generated. Although these are actually two concepually different blocks code, for convenience I combined it in one Mathematica notebook. Thereby you can change the game definition and observe the new results immediately in the same notebook.

The code that follows gives a small example what a repayment game definition looks like and also shows the corresponding output of the program. Of course the code and the results will change if the user defines own repayment games.

A.6.1 Source code (example)

```
(* ================= *)
(* GAME DEFINITION *)

(* ---- Number of Players ---- *)

n=2;

(* ---- Actionset for each player ---- *)

actionSet [1]={
{"0R-n",0,"n"},
{"0R-y",0,"y"},
{"1R",-R,"y"},
{"2R",-2R,"y"}};

actionSet [2]= actionSet [1];

(* ---- Punishment function: borrower to deadweight, controlled by lender ---- *)

punishmentFunction[ abilityCombination_ , actionCombination_, playerKey_]:=Refine[Piecewise[{
{0,          (fnGetActionMessage/@actionCombination)==ConstantArray["y",n]&& Plus@@(fnGetActionRepayment/@actionCombination)<=-n R},(* Rule (1) *)
{0,          (fnGetActionRepayment/@actionCombination)==ConstantArray[0,n]&&fnGetActionMessage[actionCombination[[playerKey]]]=="n" },(* Rule (2a) *)
{-p,         (fnGetActionRepayment/@actionCombination)==ConstantArray[0,n]&&fnGetActionMessage[actionCombination[[playerKey]]]=="y" },(* Rule (2b) *)
{-p,         fnGetActionMessage[actionCombination[[playerKey]]]=="n"&&Plus@@(fnGetActionRepayment/@actionCombination)<=-n R},(* Rule (3a) *)
{0,          Union[Delete[(fnGetActionMessage/@actionCombination),playerKey]]=={"n"}&&Union[Delete[(fnGetActionRepayment/@actionCombination),playerKey]]=={0}&&fnGetActionMessage[actionCombination
       [[playerKey]]]=="y"&&fnGetActionRepayment[actionCombination[[playerKey]]]<=-n R},(* Rule (3a) *)
{-p,         True}(* Rule (3)&(4) *)
}]];

(* ---- Transfer function: borrower to lender, controlled by lender ---- *)

rewardFunction[ abilityCombination_ , actionCombination_, playerKey_]:=Refine[Piecewise[{
{+n R+c, Union[Delete[(fnGetActionMessage/@actionCombination),playerKey]]=={"n"}&&Union[Delete[(fnGetActionRepayment/@actionCombination),playerKey]]=={0}&&fnGetActionMessage[actionCombination[[
       playerKey]]]=="y"&&fnGetActionRepayment[actionCombination[[playerKey]]]<=-n R},(* Rule (3a) *)
{0,          True}(* Rule (3)&(4) *)
}]];

(* ---- Sanctions function: borrower to deadweigh, controlled by group member ---- *)

sanctionFunction [ abilityCombination_ , actionCombination_, playerNr_]:=0;

(* ---- How the layers (apart from repayment) combine to yield the payoff. This is a game-specific information. ---- *)

payoffLayers [ abilityCombination_ , actionCombination_, playerNr_]:=+ punishmentFunction[ abilityCombination , actionCombination, playerNr] + rewardFunction[ abilityCombination , actionCombination, playerNr]+
       sanctionFunction [ abilityCombination , actionCombination, playerNr ];

(* ---- Assumptions ---- *)

$Assumptions=True; (* Hygiene: Clearing list of global assumptions first . *)

```

```
(* Caution: Use == for comparisons, do not use = instead! Watch out for contradictory assumptions! Do not use ambiguous assumptions like <=! *)
$Assumptions=And[
R> 0, (* sign gross interest *)
p>2R, (* intensity punishment *)
e>0]; (* intensity reward *)



(* ============= *)
(* GAME RESULTS *)

(* ---- Game Construction & Solution ---- *)

solutionMatrixTos=MapIndexed[fnSolveGame[game[#2],feasibilityMatrix[#2]]&,emptyTos,{n}];

(* ---- Main Results ---- *)

headline ["Map_of_equilibria"]
Map[fnGetEqActions,solutionMatrixTos,{n}];
fnDisplayTos[%]

lenderH=headline["Map_of_lender_payoff"] ;
MapIndexed[fnGetEqLenderPayoffTotal[#2,#1]&,solutionMatrixTos,{n}];
Map[fnFormatCombinationsList,%,{n}];
lender=fnDisplayTos[%];

borrowerH=headline["Map_of_borrower_payoffs\n"] ;
MapIndexed[fnGetEqPayoffs[#2,#1]&,solutionMatrixTos,{n}];
Map[fnFormatCombinationsList,%,{n}];
borrower=fnDisplayTos[%];

lossH=headline ["Map_of_deadweight_loss"];
MapIndexed[fnGetEqLoss[#2,#1]&,solutionMatrixTos,{n}];
Map[fnFormatCombinationsList,%,{n}];
loss=fnDisplayTos[%];

Grid[{{lenderH,borrowerH,lossH},{lender,borrower,loss}},Frame->All,Spacings->{5,5}] (* Combines results in a grid *)

(* ---- Further Results ---- *)

headline ["Traceable_Solution"]
MapIndexed[fnSolveGameTraceable[game[#2],feasibilityMatrix[#2]]&,emptyTos,{n}];
Map[fnFormatSolutionMatrix,%,{n}];
Map[fnDisplayMatrix,%,{n}];
fnDisplayTos[%]

headline ["Payoff_layers_(apart_from_repayment)_"]
layer=payoffLayers; (* Choose payoff layer here *)
fnGetPayoffLayerCombinationsTos[layer]/.{Null-> cInfString};
Map[fnFormatCombination[#]&,%,{2n}];
Map[fnDisplayMatrix[#]&,%,{n}];
fnDisplayTos[%]
```

```
(*
headline["Traceable Solution - A single shape"] (* Useful to check huge outputs *)
MapIndexed[fnSolveGameTraceable[game[#2],feasibilityMatrix[#2]]&,emptyTos,{n}];
%[[2,2]](* addresses the shape by its abilityKey ; make sure the shape exists ! *)
Map[fnFormatSolutionMatrix,%,{n}];
fnDisplayMatrix[%]
*)
```

A.6.2 *Software output (example)*

Punishment strategy

	P2:0R-n	P2:0R-y	P2:1R	P2:2R
P1:0R-n	0 | 0	0 | -p	-p | -p	-p | e+2R
P1:0R-y	-p | 0	-p | -p	-p | -p	0 | 0
P1:1R	-p | -p	-p | -p	0 | 0	0 | 0
P1:2R	e+2R | -p	0 | 0	0 | 0	0 | 0

Map of equilibria

	A2:0	A2:-R	A2:-2 R
A1:0	0R-n | 0R-n	0R-n | 0R-n	
A1:-R	0R-n | 0R-n	0R-n | 0R-n 1R | 1R	1R | 1R
A1:-2 R		1R | 1R	1R | 1R

A*i*:	ability to repay for borrower *i*
P*i*:	action of player *i*
×	infeasible actions

Map of lender payoff

	A2:0	A2:-R	A2:-2 R
A1:0	0	0	
A1:-R	0	0 2R	2R
A1:-2 R		2R	2R

Map of borrower payoffs

	A2:0	A2:-R	A2:-2 R
A1:0	0 | 0	0 | 0	
A1:-R	0 | 0	0 | 0 -R | -R	-R | -R
A1:-2 R		-R | -R	-R | -R

Map of deadweight loss

	A2:0	A2:-R	A2:-2 R
A1:0	0	0	
A1:-R	0	0 0	0
A1:-2 R		0	0

Traceable solution

A1:0 — A2:0

	P2:0R-n	P2:0R-y	P2:1R	P2:2R
P1:0R-n	[0,0]	[0,-p]	×	×
P1:0R-y	[-p,0]	[-p,-p]	×	×
P1:1R	×	×	×	×
P1:2R	×	×	×	×

A1:0 — A2:-R

	P2:0R-n	P2:0R-y	P2:1R	P2:2R
P1:0R-n	[0,0]	[0,-p]	[-p,-p-R]	×
P1:0R-y	[-p,0]	[-p,-p]	[-p,-p-R]	×
P1:1R	×	×	×	×
P1:2R	×	×	×	×

A1:0 — A2:-2R

	P2:0R-n	P2:0R-y	P2:1R	P2:2R
P1:0R-n	[0,0]	[0,-p]	[-p,-p-R]	[-p,e]
P1:0R-y	[-p,0]	[-p,-p]	[-p,-p-R]	[0,-2 R]
P1:1R	×	×	×	×
P1:2R	×	×	×	×

A1:-R — A2:0

	P2:0R-n	P2:0R-y	P2:1R	P2:2R
P1:0R-n	[0,0]	[0,-p]	×	×
P1:0R-y	[-p,0]	[-p,-p]	×	×
P1:1R	[-p-R,-p]	[-p-R,-p]	×	×
P1:2R	×	×	×	×

A1:-R — A2:-R

	P2:0R-n	P2:0R-y	P2:1R	P2:2R
P1:0R-n	[0,0]	[0,-p]	[-p,-p-R]	×
P1:0R-y	[-p,0]	[-p,-p]	[-p,-p-R]	×
P1:1R	[-p-R,-p]	[-p-R,-p]	[-R,-R]	×
P1:2R	×	×	×	×

A1:-R — A2:-2R

	P2:0R-n	P2:0R-y	P2:1R	P2:2R
P1:0R-n	[0,0]	[0,-p]	[-p,-p-R]	[-p,e]
P1:0R-y	[-p,0]	[-p,-p]	[-p,-p-R]	[0,-2 R]
P1:1R	[-p-R,-p]	[-p-R,-p]	[-R,-R]	[-R,-2 R]
P1:2R	×	×	×	×

A1:-2R — A2:0

	P2:0R-n	P2:0R-y	P2:1R	P2:2R
P1:0R-n	[0,0]	[0,-p]	×	×
P1:0R-y	[-p,0]	[-p,-p]	×	×
P1:1R	[-p-R,-p]	[-p-R,-p]	×	×
P1:2R	[e,-p]	[-2 R,0]	×	×

A1:-2R — A2:-R

	P2:0R-n	P2:0R-y	P2:1R	P2:2R
P1:0R-n	[0,0]	[0,-p]	[-p,-p-R]	×
P1:0R-y	[-p,0]	[-p,-p]	[-p,-p-R]	×
P1:1R	[-p-R,-p]	[-p-R,-p]	[-R,-R]	×
P1:2R	[e,-p]	[-2 R,0]	[-2 R,-R]	×

A1:-2R — A2:-2R

	P2:0R-n	P2:0R-y	P2:1R	P2:2R
P1:0R-n	[0,0]	[0,-p]	[-p,-p-R]	[-p,e]
P1:0R-y	[-p,0]	[-p,-p]	[-p,-p-R]	[0,-2 R]
P1:1R	[-p-R,-p]	[-p-R,-p]	[-R,-R]	[-R,-2 R]
P1:2R	[e,-p]	[-2 R,0]	[-2 R,-R]	[-2 R,-2 R]

A.7 Potential improvements

Note that the code presented in this dissertation is prototype code not industry code. The following sections therefore show the directions in which further improvements can be made.

A.7.1 Functionality

The flows of value (p. 103) in the repayment game should be implemented in a conceptually cleaner way. A general 'flow-function' should capture origin, target and value of a flow. This would also require a thought-through naming convention. Finally, the individual flows must be assembled to the players' payoff functions.

It would be nice to have a function that calculates avoidable harm (see 11) automatically. It would have to somehow compare the second player's payoff under various actions of the first player – and vice versa. Caution: In a game with three players the person responsible for causing avoidable harm becomes ambiguous.

The scope of application could be extended by including more solution concepts. A consideration could be given to sequentially played games, continuous games, refined game theoretic equilibrium concepts. Maybe even behavioural rules could be implemented to improve the match between theory and reality.

I also have the feeling that with more expertise in informatics, numerics and algorithm theory the performance of the code could be improved with respect to CPU and memory consumption. This quickly becomes an issue when solving 'bigger' games. Especially solving games with more than three players takes a considerable amount of time. So recycling and reusing every calculation is important. In symmetric games, for example, symmetric shapes should be calculated only once and the solution-matrix should simply be transposed for the other cases.

A.7.2 Usability

A pressing issue to make the algorithm accessible to a wider audience of users would be to increase its usability. This includes a user interface for easy input without the need for Mathematica knowledge. It would also include plausibility checks of the user input. For example contradicting assumptions should be detected right away, before executing the algorithm to solve the game. Likewise whether every borrower has a well-defined action set should be checked beforehand. It would also include a nicely formatted output and some export functions. When I copied the game results to the LaTeX-file of this dissertation, I used Right-Click "Copy as LaTeX" in the Mathematica context menu. After inserting the string into LaTeX each time I performed numerous replacements to produce nicely formatted tables.

For map of equilibria/payoffs/deadweight loss		**For traceable solution**	
Replace …	by …	Replace …	by …
`$\|$`	`\textbar\`	`array`	`tabular`
`\fbox`	`\makecell`	`\fbox{$`	`{`
`$\text{`		`$}`	`}`
`$}`		`$\times$`	`\texttimes\`
`$0`	`0}`	`\text{}`	
`$\backslash $n`	`\\`		
`\text{`	`\thead{`		

Thereafter I had to manually insert some horizontal lines and put everything in an `lstlistings` environment.

Actually, the whole process could be automated using Mathematica's extensive collection of export functions like `ExportString[%,"LaTeXFragment"]` or like the command `Export["file.tex,%,"LaTeX"]`. By using the commands `StringReplace` or `ConversionRules` the necessary replacements could be defined. By `SetDirectory[]` the location where the exported file is saved can be specified, such that the ready prepared file just has to be inserted into LaTeX using the latex command `\input{file}`.

A.7.3 Code structure

The code should be structured better making use of Mathematica packages. A prerequisite to do this is a thorough analysis of which functions can be private and which methods should be public. The defined Mathematica packages could then simply be loaded at the beginning of the code and the user would not need to manually load the individual code blocks (p. 154) into the Mathematica kernel.

A.7.4 Documentation

Finally, the documentation of the code could be improved. It would be particularly helpful to draw some charts that illuminate the dependencies between the functions of the algorithm. Maybe a dataflow-diagram as used in flow-based programming would be adequate.

B Detailed solutions of repayment games

C. Prem, *The Theory of Credit Contracts*,
https://doi.org/10.1007/978-3-658-29362-8

B.1 Individual liability

```
(* GAME DEFINITION – Individual liability *)

(* NUMBER OF PLAYERS ---- *)
n=2;

(* ACTIONSET for each player ---- *)
actionSet[1]={
{"0R",0,""},
{"1R",-R,""},
{"2R",-2R,""}};
actionSet[2]=actionSet[1];

(* PUNISHMENT FUNCTION: borrower to deadweight, controlled by lender ---- *)
punishmentFunction[abilityCombination_, actionCombination_, playerKey_]:=Refine[Piecewise[{
{-M,      -fnGetActionRepayment[actionCombination[[playerKey]]]<R (* punishment if default ... *)
&& -Plus@@(fnGetActionRepayment/@actionCombination)< 2 R (* ... and peer did not kindly pay for him *)},
{0,             True}(* No punishment otherwise *)
}]];

(* REWARD FUNCTION: lender to borrower flow ---- *)
rewardFunction[abilityCombination_, actionCombination_, playerKey_]:=0;

(* SANCTIONS FUNCTION: borrower to deadweight flow ---- *)
sanctionFunction[abilityCombination_, actionCombination_, playerKey_]:=0;

(* ASSEMBLING PAYOFFLAYERS (apart from repayment) ---- *)
payoffLayers[abilityCombination_, actionCombination_, playerNr_]:=+ punishmentFunction[abilityCombination, actionCombination, playerNr] + rewardFunction[abilityCombination, actionCombination, playerNr]+
        sanctionFunction[abilityCombination, actionCombination, playerNr];

(* ASSUMPTIONS ---- *)
$Assumptions=True; (* Hygiene: Clearing list of global assumptions first . *)
$Assumptions=And[
R> 0, (* sign gross interest *)
M>R]; (* intensity punishment *)
(* Use == for comparisons! No contradictory assumptions! No ambiguous assumptions like <=! *)
```

INDIVIDUAL LIABILITY

Medium punishment intensity
($R < M$)

Punishment strategy

	P2:0R	P2:1R	P2:2R
P1:0R	-M | -M	-M | 0	0 | 0
P1:1R	0 | -M	0 | 0	0 | 0
P1:2R	0 | 0	0 | 0	0 | 0

A*i*:	ability to repay for borrower *i*
P*i*:	action of player *i*
×	infeasible actions

Map of equilibria

	A2:0	A2:-R	A2:-2 R
A1:0	0R | 0R	0R | 1R	0R | 1R
A1:-R	1R | 0R	1R | 1R	1R | 1R
A1:-2 R	1R | 0R	1R | 1R	1R | 1R

Map of lender payoffs

	A2:0	A2:-R	A2:-2 R
A1:0	0	R	R
A1:-R	R	2R	2R
A1:-2 R	R	2R	2R

Map of deadweight loss

	A2:0	A2:-R	A2:-2 R
A1:0	-2M	-M	-M
A1:-R	-M	0	0
A1:-2 R	-M	0	0

Traceable solution

		A2:0			A2:-R			A2:-2 R		
		P2:0R	P2:1R	P2:2R	P2:0R	P2:1R	P2:2R	P2:0R	P2:1R	P2:2R
A1:0	P1:0R	[-M,-M]	×	×	[-M,-M]	[-M,-R]	×	[-M,-M]	[-M,-R]	[0,-2 R]
	P1:1R	×	×	×	×	×	×	×	×	×
	P1:2R	×	×	×	×	×	×	×	×	×
		P2:0R	P2:1R	P2:2R	P2:0R	P2:1R	P2:2R	P2:0R	P2:1R	P2:2R
A1:-R	P1:0R	[-M,-M]	×	×	[-M,-M]	[-M,-R]	×	[-M,-M]	[-M,-R]	[0,-2 R]
	P1:1R	[-R,-M]	×	×	[-R,-M]	[-R,-R]	×	[-R,-M]	[-R,-R]	[-R,-2 R]
	P1:2R	×	×	×	×	×	×	×	×	×
		P2:0R	P2:1R	P2:2R	P2:0R	P2:1R	P2:2R	P2:0R	P2:1R	P2:2R
A1:-2 R	P1:0R	[-M,-M]	×	×	[-M,-M]	[-M,-R]	×	[-M,-M]	[-M,-R]	[0,-2 R]
	P1:1R	[-R,-M]	×	×	[-R,-M]	[-R,-R]	×	[-R,-M]	[-R,-R]	[-R,-2 R]
	P1:2R	[-2 R,0]	×	×	[-2 R,0]	[-2 R,-R]	×	[-2 R,0]	[-2 R,-R]	[-2 R,-2 R]

B.2 Joint liability – High punishment intensity

```
(* GAME DEFINITION – Joint liability with high punishment intensity *)

(* NUMBER OF PLAYERS ---- *)
n=2;

(* ACTIONSET for each player ---- *)
actionSet [1]={
{"0R",0,""},
{"1R",−R,""},
{"2R",−2R,""}};
actionSet [2]= actionSet [1];

(* PUNISHMENT FUNCTION: borrower to deadweight, controlled by lender flow ---- *)
punishmentFunction[ abilityCombination_ , actionCombination_, playerKey_]:=Refine[Piecewise[{
{−P,        Plus@@(fnGetActionRepayment/@actionCombination)>−2R},(* Punishment if group defaults *)
{0,             True}(* No punishment otherwise *)
}]];
(* Order of conditions matters! First that yields TRUE is taken *)

(* REWARD FUNCTION: lender to borrower flow ---- *)
rewardFunction[ abilityCombination_ , actionCombination_, playerKey_] := 0;

(* SANCTIONS FUNCTION: borrower to deadweight flow ---- *)
sanctionFunction [ abilityCombination_ , actionCombination_, playerKey_] := 0;

(* ASSEMBLING PAYOFFLAYERS (apart from repayment) ---- *)
payoffLayers [ abilityCombination_ , actionCombination_, playerNr_] :=
+ punishmentFunction[ abilityCombination , actionCombination, playerNr]
+ rewardFunction[ abilityCombination , actionCombination, playerNr]
+ sanctionFunction [ abilityCombination , actionCombination, playerNr ];

(* ASSUMPTIONS ---- *)
$Assumptions=True; (* Hygiene: Clearing list of global assumptions first . *)
$Assumptions=And[
R> 0, (* sign gross interest *)
P>2R]; (* intensity punishment *)
(* Use == for comparisons! No contradictory assumptions! No ambiguous assumptions like <=! *)
```

JOINT LIABILITY

High punishment intensity
($2R < P$)

Punishment strategy

	P2:0R	P2:1R	P2:2R
P1:0R	-P | -P	-P | -P	0 | 0
P1:1R	-P | -P	0 | 0	0 | 0
P1:2R	0 | 0	0 | 0	0 | 0

A*i*:	ability to repay for borrower *i*
P*i*:	action of player *i*
×	infeasible actions

Map of equilibria

	A2:0	A2:-R	A2:-2 R
A1:0	0R | 0R	0R | 0R	0R | 2R
A1:-R	0R | 0R	0R | 0R 1R | 1R	0R | 2R 1R | 1R
A1:-2 R	2R | 0R	1R | 1R 2R | 0R	0R | 2R 1R | 1R 2R | 0R

Map of lender payoffs

	A2:0	A2:-R	A2:-2 R
A1:0	0	0	2R
A1:-R	0	0 2R	2R 2R
A1:-2 R	2R	2R 2R	2R 2R 2R

Map of deadweight loss

	A2:0	A2:-R	A2:-2 R
A1:0	-2P	-2P	0
A1:-R	-2P	-2P 0	0 0
A1:-2 R	0	0 0	0 0 0

Traceable solution

		A2:0			A2:-R			A2:-2 R		
		P2:0R	P2:1R	P2:2R	P2:0R	P2:1R	P2:2R	P2:0R	P2:1R	P2:2R
A1:0	P1:0R	[-P,-P]	×	×	[-P,-P]	[-P,-P-R]	×	[-P,-P]	[-P,-P-R]	[0,-2 R]
	P1:1R	×	×	×	×	×	×	×	×	×
	P1:2R	×	×	×	×	×	×	×	×	×
A1:-R	P1:0R	[-P,-P]	×	×	[-P,-P]	[-P,-P-R]	×	[-P,-P]	[-P,-P-R]	[0,-2 R]
	P1:1R	[-P-R,-P]	×	×	[-P-R,-P]	[-R,-R]	×	[-P-R,-P]	[-R,-R]	[-R,-2 R]
	P1:2R	×	×	×	×	×	×	×	×	×
A1:-2 R	P1:0R	[-P,-P]	×	×	[-P,-P]	[-P,-P-R]	×	[-P,-P]	[-P,-P-R]	[0,-2 R]
	P1:1R	[-P-R,-P]	×	×	[-P-R,-P]	[-R,-R]	×	[-P-R,-P]	[-R,-R]	[-R,-2 R]
	P1:2R	[-2 R,0]	×	×	[-2 R,0]	[-2 R,-R]	×	[-2 R,0]	[-2 R,-R]	[-2 R,-2 R]

B.3 Joint liability – Adaptive punishment intensity

```
(* GAME DEFINITION – Joint liability with adaptive punishment intensity *)

(* NUMBER OF PLAYERS----- *)
n=2;

(* ACTIONSET for each player----- *)
actionSet[1]={
{"0R",0,""},
{"1R",-R,""},
{"2R",-2R,""}};
actionSet[2]=actionSet[1];

(* PUNISHMENT FUNCTION: borrower to deadweight flow ----- *)
punishmentFunction[abilityCombination_,actionCombination_,playerKey_]:=Refine[Piecewise[{
{-P,     -Plus@@(fnGetActionRepayment/@actionCombination)==0},(* Punishment if no repayment *)
{-M,     -Plus@@(fnGetActionRepayment/@actionCombination)<n R},(* Punishment if partial repayment *)
{0,             True}(*No punishment otherwise*)}]];
(* Order of conditions matters! First that yields TRUE is taken *)

(* REWARD FUNCTION: lender to borrower flow ----- *)
rewardFunction[abilityCombination_,actionCombination_,playerKey_]:=0;

(* SANCTIONS FUNCTION: borrower to deadweight flow ----- *)
sanctionFunction[abilityCombination_,actionCombination_,playerKey_]:=0;

(* ASSEMBLING PAYOFFLAYERS (apart from repayment) ----- *)
payoffLayers[abilityCombination_,actionCombination_,playerNr_]:=
+ punishmentFunction[abilityCombination,actionCombination,playerNr]
+ rewardFunction[abilityCombination,actionCombination,playerNr]
+ sanctionFunction[abilityCombination,actionCombination,playerNr];

(* ASSUMPTIONS----- *)
$Assumptions=True;(*Hygiene:Clearing list of global assumptions first .*)
$Assumptions=And[
R>0,(* sign gross interest *)
R<M<2R,(* intensity mild punishment *)
P>R+m];(* intensity punishment *)
(* Use == for comparisons! No contradictory assumptions! No ambiguous assumptions like <=! *)
```

Note: You can even get rid of the free-riding equilibria by adding the following punishment rule to the punishment function:

```
{-M,     -Plus @@ (fnGetActionRepayment /@ actionCombination) == n R && fnGetActionRepayment[actionCombination[[playerKey]]] == 0}
```

JOINT LIABILITY

Adaptive punishment intensity
($R < M < 2R$ and $R + M < P$)

Punishment strategy

<table>
<tr><th></th><th>P2:0R</th><th>P2:1R</th><th>P2:2R</th></tr>
<tr><td>P1:0R</td><td>-P | -P</td><td>-M | -M</td><td>0 | 0</td></tr>
<tr><td>P1:1R</td><td>-M | -M</td><td>0 | 0</td><td>0 | 0</td></tr>
<tr><td>P1:2R</td><td>0 | 0</td><td>0 | 0</td><td>0 | 0</td></tr>
</table>

A*i*:	ability to repay for borrower *i*
P*i*:	action of player *i*
×	infeasible actions

Map of equilibria

<table>
<tr><th></th><th>A2:0</th><th>A2:-R</th><th>A2:-2 R</th></tr>
<tr><td>A1:0</td><td>0R | 0R</td><td>0R | 1R</td><td>0R | 2R</td></tr>
<tr><td>A1:-R</td><td>1R | 0R</td><td>1R | 1R</td><td>0R | 2R
1R | 1R</td></tr>
<tr><td>A1:-2 R</td><td>2R | 0R</td><td>1R | 1R
2R | 0R</td><td>0R | 2R
1R | 1R
2R | 0R</td></tr>
</table>

Map of lender payoffs

	A2:0	A2:-R	A2:-2 R
A1:0	0	R	2R
A1:-R	R	2R	2R 2R
A1:-2 R	2R	2R 2R	2R 2R 2R

Map of deadweight loss

	A2:0	A2:-R	A2:-2 R
A1:0	-2P	-2M	0
A1:-R	-2M	0	0 0
A1:-2 R	0	0 0	0 0 0

Traceable solution

A1:0, A2:0

	P2:0R	P2:1R	P2:2R
P1:0R	[-P,-P]	×	×
P1:1R	×	×	×
P1:2R	×	×	×

A1:0, A2:-R

	P2:0R	P2:1R	P2:2R
P1:0R	[-P,-P]	[-M,-M-R]	×
P1:1R	×	×	×
P1:2R	×	×	×

A1:0, A2:-2 R

	P2:0R	P2:1R	P2:2R
P1:0R	[-P,-P]	[-M,-M-R]	[0,-2 R]
P1:1R	×	×	×
P1:2R	×	×	×

A1:-R, A2:0

	P2:0R	P2:1R	P2:2R
P1:0R	[-P,-P]	×	×
P1:1R	[-M-R,-M]	×	×
P1:2R	×	×	×

A1:-R, A2:-R

	P2:0R	P2:1R	P2:2R
P1:0R	[-P,-P]	[-M,-M-R]	×
P1:1R	[-M-R,-M]	[-R,-R]	×
P1:2R	×	×	×

A1:-R, A2:-2 R

	P2:0R	P2:1R	P2:2R
P1:0R	[-P,-P]	[-M,-M-R]	[0,-2 R]
P1:1R	[-M-R,-M]	[-R,-R]	[-R,-2 R]
P1:2R	×	×	×

A1:-2 R, A2:0

	P2:0R	P2:1R	P2:2R
P1:0R	[-P,-P]	×	×
P1:1R	[-M-R,-M]	×	×
P1:2R	[-2 R,0]	×	×

A1:-2 R, A2:-R

	P2:0R	P2:1R	P2:2R
P1:0R	[-P,-P]	[-M,-M-R]	×
P1:1R	[-M-R,-M]	[-R,-R]	×
P1:2R	[-2 R,0]	[-2 R,-R]	×

A1:-2 R, A2:-2 R

	P2:0R	P2:1R	P2:2R
P1:0R	[-P,-P]	[-M,-M-R]	[0,-2 R]
P1:1R	[-M-R,-M]	[-R,-R]	[-R,-2 R]
P1:2R	[-2 R,0]	[-2 R,-R]	[-2 R,-2 R]

B.4 Joint liability – High punishment intensity (with social sanctions)

```
(* GAME DEFINITION – Joint liability – high punishment intensity (with social sanctions) *)

(* NUMBER OF PLAYERS ---- *)
n=2;
(* ACTIONSET for each player ---- *)
actionSet [1]={
{"0R",0,""},
{"1R",-R,""},
{"2R",-2R,""}};
actionSet [2]= actionSet [1];

(* PUNISHMENT FUNCTION: borrower to deadweight, controlled by lender ---- *)
punishmentFunction[ abilityCombination_ , actionCombination_, playerKey_]:=Refine[Piecewise[{
{-P,         Plus@@(fnGetActionRepayment/@actionCombination)>-2R},(* Punishment if group defaults *)
{0,              True}(* No punishment otherwise *)
}]];
(* REWARD FUNCTION: lender to borrower flow ---- *)
rewardFunction[ abilityCombination_ ,actionCombination_,playerKey_]:=0;
(* SANCTIONS FUNCTION: borrower to deadweight flow ---- *)
sanctionFunction [ abilityCombination_ ,actionCombination_, playerKey_]:=Refine[Piecewise[{
{-Sn,     -Plus@@(fnGetActionRepayment/@actionCombination)<n R (* not all loans are repaid *)
&& -Plus@@abilityCombination>=n R (* full repayment is possible *)
&& -fnGetActionRepayment[actionCombination[[playerKey]]]<-abilityCombination[[playerKey]](* player still has resources *)}, (* not-helping condition *)
{-Sf,   -Plus@@(fnGetActionRepayment/@actionCombination)>= n R (* all loans are repaid *)
&& -abilityCombination[[playerKey]]>=R (* can repay his loan *)
&& fnGetActionRepayment[actionCombination[[playerKey]]]==0 (* does not repay his loan *)}, (* free-riding condition *)
{0,True}
}]];

(* ASSEMBLING PAYOFFLAYERS (apart from repayment) ---- *)
payoffLayers [ abilityCombination_ ,actionCombination_, playerNr_]:=+ punishmentFunction[ abilityCombination ,actionCombination, playerNr] + rewardFunction[ abilityCombination ,actionCombination, playerNr]+
        sanctionFunction [ abilityCombination ,actionCombination, playerNr ];

(* ASSUMPTIONS ---- *)
$Assumptions=True; (* Hygiene: Clearing list of global assumptions first . *)
$Assumptions=And[
R> 0, (* sign gross interest *)
Sn>(n-1)R,(* intensity not-helping sanctions *)
Sf>R, (* intensity free-riding sanctions *)
P>n R]; (* intensity punishment *)
(* Use ==for comparisons! No contradictory assumptions! No ambiguous assumptions like <=! *)
```

JOINT LIABILITY

(with social sanctions)
High punishment intensity
($2R < P$)

Punishment strategy

	P2:0R	P2:1R	P2:2R
P1:0R	-P | -P	-P | -P	0 | 0
P1:1R	-P | -P	0 | 0	0 | 0
P1:2R	0 | 0	0 | 0	0 | 0

Ai:	ability to repay for borrower *i*
Pi:	action of player *i*
×	infeasible actions

Map of equilibria

	A2:0	A2:-R	A2:-2 R
A1:0	0R | 0R	0R | 0R	0R | 2R
A1:-R	0R | 0R	1R | 1R	1R | 1R
A1:-2 R	2R | 0R	1R | 1R	1R | 1R

Map of lender payoffs

	A2:0	A2:-R	A2:-2 R
A1:0	0	0	2R
A1:-R	0	2R	2R
A1:-2 R	2R	2R	2R

Map of deadweight loss

	A2:0	A2:-R	A2:-2 R
A1:0	-2P	-2P	0
A1:-R	-2P	0	0
A1:-2 R	0	0	0

Traceable solution

		A2:0			A2:-R			A2:-2 R		
		P2:0R	P2:1R	P2:2R	P2:0R	P2:1R	P2:2R	P2:0R	P2:1R	P2:2R
A1:0	P1:0R	[-P,-P]	×	×	[-P,-P]	[-P,-P-R]	×	[-P,-P-Sn]	[-P,-P-R-Sn]	[0,-2 R]
	P1:1R	×	×	×	×	×	×	×	×	×
	P1:2R	×	×	×	×	×	×	×	×	×
A1:-R	P1:0R	[-P,-P]	×	×	[-P-Sn,-P-Sn]	[-P-Sn,-P-R]	×	[-P-Sn,-P-Sn]	[-P-Sn,-P-R-Sn]	[-Sf,-2 R]
	P1:1R	[-P-R,-P]	×	×	[-P-R,-P-Sn]	[-R,-R]	×	[-P-R,-P-Sn]	[-R,-R]	[-R,-2 R]
	P1:2R	×	×	×	×	×	×	×	×	×
A1:-2 R	P1:0R	[-P-Sn,-P]	×	×	[-P-Sn,-P-Sn]	[-P-Sn,-P-R]	×	[-P-Sn,-P-Sn]	[-P-Sn,-P-R-Sn]	[-Sf,-2 R]
	P1:1R	[-P-R-Sn,-P]	×	×	[-P-R-Sn,-P-Sn]	[-R,-R]	×	[-P-R-Sn,-P-Sn]	[-R,-R]	[-R,-2 R]
	P1:2R	[-2 R,0]	×	×	[-2 R,-Sf]	[-2 R,-R]	×	[-2 R,-Sf]	[-2 R,-R]	[-2 R,-2 R]

B.5 Joint liability – Medium punishment intensity

```
(* GAME DEFINITION – Joint liability with medium punishment intensity *)

(* NUMBER OF PLAYERS ---- *)
n=2;

(* ACTIONSET for each player ---- *)
actionSet[1]={
{"0R",0,""},
{"1R",-R,""},
{"2R",-2R,""}};
actionSet[2]=actionSet[1];

(* PUNISHMENT FUNCTION: borrower to deadweight, controlled by lender flow ---- *)
punishmentFunction[abilityCombination_,actionCombination_,playerKey_]:=Refine[Piecewise[{
{-M,        Plus@@(fnGetActionRepayment/@actionCombination)>-2R},(* Punishment if default *)
{0,              True}(* No punishment otherwise *)
}]];
(* Order of conditions matters! First that yields TRUE is taken *)

(* REWARD FUNCTION: lender to borrower flow ---- *)
rewardFunction[abilityCombination_, actionCombination_,playerKey_] := 0;

(* SANCTIONS FUNCTION: borrower to deadweight flow ---- *)
sanctionFunction[abilityCombination_, actionCombination_,playerKey_] := 0;

(* ASSEMBLING PAYOFFLAYERS (apart from repayment) ---- *)
payoffLayers[abilityCombination_, actionCombination_,playerNr_] :=
+ punishmentFunction[abilityCombination, actionCombination,playerNr]
+ rewardFunction[abilityCombination, actionCombination, playerNr]
+ sanctionFunction[abilityCombination, actionCombination, playerNr];

(* ASSUMPTIONS ---- *)
$Assumptions=True; (* Hygiene: Clearing list of global assumptions first. *)
$Assumptions=And[
R>0,(* sign gross interest *)
R<M<2R];(* intensity punishment *)
(* Use == for comparisons! No contradictory assumptions! No ambiguous assumptions like <=! *)
```

JOINT LIABILITY

Medium punishment intensity
$(R < M < 2R)$

Punishment strategy

	P2:0R	P2:1R	P2:2R
P1:0R	-M | -M	-M | -M	0 | 0
P1:1R	-M | -M	0 | 0	0 | 0
P1:2R	0 | 0	0 | 0	0 | 0

Ai:	ability to repay for borrower *i*
Pi:	action of player *i*
×	infeasible actions

Map of equilibria

	A2:0	A2:-R	A2:-2 R
A1:0	0R | 0R	0R | 0R	0R | 0R
A1:-R	0R | 0R	0R | 0R 1R | 1R	0R | 0R 1R | 1R
A1:-2 R	0R | 0R	0R | 0R 1R | 1R	0R | 0R 1R | 1R

Map of lender payoffs

	A2:0	A2:-R	A2:-2 R
A1:0	0	0	0
A1:-R	0	0 2R	0 2R
A1:-2 R	0	0 2R	0 2R

Map of deadweight loss

	A2:0	A2:-R	A2:-2 R
A1:0	-2M	-2M	-2M
A1:-R	-2M	-2M 0	-2M 0
A1:-2 R	-2M	-2M 0	-2M 0

Traceable solution

A2:0

		P2:0R	P2:1R	P2:2R
A1:0	P1:0R	[-M,-M]	×	×
	P1:1R	×	×	×
	P1:2R	×	×	×
A1:-R	P1:0R	[-M,-M]	×	×
	P1:1R	[-M-R,-M]	×	×
	P1:2R	×	×	×
A1:-2 R	P1:0R	[-M,-M]	×	×
	P1:1R	[-M-R,-M]	×	×
	P1:2R	[-2 R,0]	×	×

A2:-R

		P2:0R	P2:1R	P2:2R
A1:0	P1:0R	[-M,-M]	[-M,-M-R]	×
	P1:1R	×	×	×
	P1:2R	×	×	×
A1:-R	P1:0R	[-M,-M]	[-M,-M-R]	×
	P1:1R	[-M-R,-M]	[-R,-R]	×
	P1:2R	×	×	×
A1:-2 R	P1:0R	[-M,-M]	[-M,-M-R]	×
	P1:1R	[-M-R,-M]	[-R,-R]	×
	P1:2R	[-2 R,0]	[-2 R,-R]	×

A2:-2 R

		P2:0R	P2:1R	P2:2R
A1:0	P1:0R	[-M,-M]	[-M,-M-R]	[0,-2 R]
	P1:1R	×	×	×
	P1:2R	×	×	×
A1:-R	P1:0R	[-M,-M]	[-M,-M-R]	[0,-2 R]
	P1:1R	[-M-R,-M]	[-R,-R]	[-R,-2 R]
	P1:2R	×	×	×
A1:-2 R	P1:0R	[-M,-M]	[-M,-M-R]	[0,-2 R]
	P1:1R	[-M-R,-M]	[-R,-R]	[-R,-2 R]
	P1:2R	[-2 R,0]	[-2 R,-R]	[-2 R,-2 R]

B.6 Joint liability – Medium punishment (with social sanctions)

```
(* GAME DEFINITION – Joint liability with medium punishment intensity (with social sanctions) *)

(* NUMBER OF PLAYERS ---- *)
n=2;
(* ACTIONSET for each player ---- *)
actionSet [1]={
{"0R",0,""},
{"1R",-R,""},
{"2R",-2R,""}};
actionSet [2]= actionSet [1];

(* PUNISHMENT FUNCTION: borrower to deadweight, controlled by lender ---- *)
punishmentFunction[ abilityCombination_ , actionCombination_, playerKey_]:=Refine[Piecewise[{
{-M,        Plus@@(fnGetActionRepayment/@actionCombination)>-2R},(* Punishment if group defaults *)
{0,              True}(* No punishment otherwise *)
}]];
(* REWARD FUNCTION: lender to borrower flow ---- *)
rewardFunction[ abilityCombination_ , actionCombination_, playerKey_]:=0;
(* SANCTIONS FUNCTION: borrower to deadweight flow ---- *)
sanctionFunction [ abilityCombination_ , actionCombination_, playerKey_]:=Refine[Piecewise[{
{-Sn,     -Plus@@(fnGetActionRepayment/@actionCombination)<n R (* not all loans are repaid *)
&& -Plus@@abilityCombination>=n R (* full repayment is possible *)
&& -fnGetActionRepayment[actionCombination[[playerKey]]]<-abilityCombination[[playerKey]](* player still has resources *)}, (* not-helping condition *)
{-Sf,   -Plus@@(fnGetActionRepayment/@actionCombination)>= n R (* all loans are repaid *)
&& -abilityCombination[[playerKey]]>=R (* can repay his loan *)
&& fnGetActionRepayment[actionCombination[[playerKey]]]==0 (* does not repay his loan *)}, (* free-riding condition *)
{0,True}
}]];

(* ASSEMBLING PAYOFFLAYERS (apart from repayment) ---- *)
payoffLayers [ abilityCombination_ , actionCombination_, playerNr_]:=+ punishmentFunction[ abilityCombination ,actionCombination, playerNr] + rewardFunction[ abilityCombination ,actionCombination, playerNr]+
          sanctionFunction [ abilityCombination ,actionCombination, playerNr ];

(* ASSUMPTIONS ---- *)
$Assumptions=True; (* Hygiene: Clearing list of global assumptions first . *)
$Assumptions=And[
R> 0, (* sign gross interest *)
Sn>(n-1)R,(* intensity not-helping sanctions *)
Sf>R, (* intensity free-riding sanctions *)
R<M<2R]; (* intensity punishment *)
(* Use ==for comparisons! No contradictory assumptions! No ambiguous assumptions like <=! *)
```

JOINT LIABILITY

(with social sanctions)
Medium punishment intensity
($R < M < 2R$)

Punishment strategy

	P2:0R	P2:1R	P2:2R
P1:0R	-M | -M	-M | -M	0 | 0
P1:1R	-M | -M	0 | 0	0 | 0
P1:2R	0 | 0	0 | 0	0 | 0

Ai:	ability to repay for borrower *i*
Pi:	action of player *i*
×	infeasible actions

Map of equilibria

	A2:0	A2:-R	A2:-2 R
A1:0	0R | 0R	0R | 0R	0R | 2R
A1:-R	0R | 0R	1R | 1R	1R | 1R
A1:-2 R	2R | 0R	1R | 1R	1R | 1R

Map of lender payoffs

	A2:0	A2:-R	A2:-2 R
A1:0	0	0	2R
A1:-R0	0	2R	2R
A1:-2 R	2R	2R	2R

Map of deadweight loss

	A2:0	A2:-R	A2:-2 R
A1:0	-2M	-2M	0
A1:-R	-2M	0	0
A1:-2 R	0	0	0

Traceable solution

		A2:0			A2:-R			A2:-2 R		
		P2:0R	P2:1R	P2:2R	P2:0R	P2:1R	P2:2R	P2:0R	P2:1R	P2:2R
A1:0	P1:0R	[-M,-M]	×	×	[-M,-M]	[-M,-M-R]	×	[-M,-M-Sn]	[-M,-M-R-Sn]	[0,-2 R]
	P1:1R	×	×	×	×	×	×	×	×	×
	P1:2R	×	×	×	×	×	×	×	×	×
A1:-R	P1:0R	[-M,-M]	×	×	[-M-Sn,-M-Sn]	[-M-Sn,-M-R]	×	[-M-Sn,-M-Sn]	[-M-Sn,-M-R-Sn]	[-Sf,-2 R]
	P1:1R	[-M-R,-M]	×	×	[-M-R,-M-Sn]	[-R,-R]	×	[-M-R,-M-Sn]	[-R,-R]	[-R,-2 R]
	P1:2R	×	×	×	×	×	×	×	×	×
A1:-2 R	P1:0R	[-M-Sn,-M]	×	×	[-M-Sn,-M-Sn]	[-M-Sn,-M-R]	×	[-M-Sn,-M-Sn]	[-M-Sn,-M-R-Sn]	[-Sf,-2 R]
	P1:1R	[-M-R-Sn,-M]	×	×	[-M-R-Sn,-M-Sn]	[-R,-R]	×	[-M-R-Sn,-M-Sn]	[-R,-R]	[-R,-2 R]
	P1:2R	[-2 R,0]	×	×	[-2 R,-Sf]	[-2 R,-R]	×	[-2 R,-Sf]	[-2 R,-R]	[-2 R,-2 R]

B.7 Joint liability – Medium punishment (with social sanctions) [n=3]

```
(* GAME DEFINITION – Joint liability with medium punishment intensity (with social sanctions) – three borrowers *)

(* NUMBER OF PLAYERS ---- *)
n=3;
(* ACTIONSET for each player ---- *)
actionSet [1]={
{"0R",0,""},
{"1R",-R,""},
{"2R",-2R,""} };
actionSet [2]= actionSet [1];
actionSet [3]= actionSet [1];

(* PUNISHMENT FUNCTION: borrower to deadweight, controlled by lender ---- *)
punishmentFunction[ abilityCombination_ ,actionCombination_ ,playerKey_]:=Refine[Piecewise[ {
{-M,        Plus@@(fnGetActionRepayment/@actionCombination)>-n R},(* Punishment if group defaults *)
{0,              True}(* No punishment otherwise *)
}]];
(* REWARD FUNCTION: lender to borrower flow ---- *)
rewardFunction[ abilityCombination_ ,actionCombination_ ,playerKey_]:=0;
(* SANCTIONS FUNCTION: borrower to deadweight flow ---- *)
sanctionFunction [ abilityCombination_ ,actionCombination_ ,playerKey_]:=Refine[Piecewise[ {
{-Sn,     -Plus@@(fnGetActionRepayment/@actionCombination)<n R (* not all loans are repaid *)
&& -Plus@@abilityCombination>=n R (* full repayment is possible *)
&& -fnGetActionRepayment[actionCombination[[playerKey]]]<-abilityCombination[[playerKey]](* player still has resources *)}, (* not-helping condition *)
{-Sf,  -Plus@@(fnGetActionRepayment/@actionCombination)>= n R (* all loans are repaid *)
&& -abilityCombination[[playerKey]]>=R (* can repay his loan *)
&& fnGetActionRepayment[actionCombination[[playerKey]]]==0 (* does not repay his loan *)}, (* free-riding condition *)
{0,True}
}]];
(* ASSEMBLING PAYOFFLAYERS (apart from repayment) ---- *)
payoffLayers [ abilityCombination_ ,actionCombination_ ,playerNr_]:=+ punishmentFunction[ abilityCombination ,actionCombination ,playerNr] + rewardFunction[ abilityCombination ,actionCombination ,playerNr]+
          sanctionFunction [ abilityCombination ,actionCombination ,playerNr ];

(* ASSUMPTIONS ---- *)
$Assumptions=True; (* Hygiene: Clearing list of global assumptions first . *)
$Assumptions=And[
R> 0, (* sign gross interest *)
Sn>(n-1)R,(* intensity not-helping sanctions *)
Sf>R, (* intensity free-riding sanctions *)
R<M<2R]; (* intensity punishment *)
(* Use == for comparisons! No contradictory assumptions! No ambiguous assumptions like <=! *)
```

JOINT LIABILITY [n=3]

(with social sanctions)
Medium punishment intensity
($R < M < 2R$)

Punishment strategy

P1:0R

<table>
<tr><th></th><th>P3:0R</th><th>P3:1R</th><th>P3:2R</th></tr>
<tr><td>P2:0R</td><td>-M | -M | -M</td><td>-M | -M | -M</td><td>-M | -M | -M</td></tr>
<tr><td>P2:1R</td><td>-M | -M | -M</td><td>-M | -M | -M</td><td>0 | 0 | 0</td></tr>
<tr><td>P2:2R</td><td>-M | -M | -M</td><td>0 | 0 | 0</td><td>0 | 0 | 0</td></tr>
</table>

P1:1R

<table>
<tr><th></th><th>P3:0R</th><th>P3:1R</th><th>P3:2R</th></tr>
<tr><td>P2:0R</td><td>-M | -M | -M</td><td>-M | -M | -M</td><td>0 | 0 | 0</td></tr>
<tr><td>P2:1R</td><td>-M | -M | -M</td><td>0 | 0 | 0</td><td>0 | 0 | 0</td></tr>
<tr><td>P2:2R</td><td>0 | 0 | 0</td><td>0 | 0 | 0</td><td>0 | 0 | 0</td></tr>
</table>

P1:2R

<table>
<tr><th></th><th>P3:0R</th><th>P3:1R</th><th>P3:2R</th></tr>
<tr><td>P2:0R</td><td>-M | -M | -M</td><td>0 | 0 | 0</td><td>0 | 0 | 0</td></tr>
<tr><td>P2:1R</td><td>0 | 0 | 0</td><td>0 | 0 | 0</td><td>0 | 0 | 0</td></tr>
<tr><td>P2:2R</td><td>0 | 0 | 0</td><td>0 | 0 | 0</td><td>0 | 0 | 0</td></tr>
</table>

Map of equilibria

<table>
<tr><th></th><th colspan="3">A1:0</th><th colspan="3">A1:-R</th><th colspan="3">A1:-2 R</th></tr>
<tr><th></th><th>A3:0</th><th>A3:-R</th><th>A3:-2 R</th><th>A3:0</th><th>A3:-R</th><th>A3:-2 R</th><th>A3:0</th><th>A3:-R</th><th>A3:-2 R</th></tr>
<tr><td>A2:0</td><td>0R | 0R | 0R</td><td>0R | 0R | 0R</td><td>0R | 0R | 0R</td><td>0R | 0R | 0R</td><td>0R | 0R | 0R</td><td>1R | 0R | 2R</td><td>0R | 0R | 0R</td><td>2R | 0R | 1R</td><td>1R | 0R | 2R
2R | 0R | 1R</td></tr>
<tr><td>A2:-R</td><td>0R | 0R | 0R</td><td>0R | 0R | 0R</td><td>0R | 1R | 2R</td><td>0R | 0R | 0R</td><td>1R | 1R | 1R</td><td>1R | 1R | 1R</td><td>2R | 1R | 0R</td><td>1R | 1R | 1R</td><td>1R | 1R | 1R</td></tr>
<tr><td>A2:-2 R</td><td>0R | 0R | 0R</td><td>0R | 2R | 1R</td><td>0R | 1R | 2R
0R | 2R | 1R</td><td>1R | 2R | 0R</td><td>1R | 1R | 1R</td><td>1R | 1R | 1R</td><td>1R | 2R | 0R
2R | 1R | 0R</td><td>1R | 1R | 1R</td><td>1R | 1R | 1R</td></tr>
</table>

Traceable solution has 3^3 shapes of dimension 3^3 and is too big to draw on DIN A4 paper.

B.8 Individual liability (with social sanctions)

```
(* GAME DEFINITION – Individual liability (with social sanctions) *)

(* NUMBER OF PLAYERS ---- *)
n=2;
(* ACTIONSET for each player ---- *)
actionSet[1]={
{"0R",0,""},
{"1R",-R,""},
{"2R",-2R,""}};
actionSet[2]=actionSet[1];

(* PUNISHMENT FUNCTION: borrower to deadweight, controlled by lender ---- *)
punishmentFunction[abilityCombination_, actionCombination_, playerKey_]:=Refine[Piecewise[{
{-M,      -fnGetActionRepayment[actionCombination[[playerKey]]]<R (* punishment if default ... *)
&& -Plus@@(fnGetActionRepayment/@actionCombination)< 2 R (* ... and peer did not kindly pay for him *)},
{0,          True}(* No punishment otherwise *)
}]];

(* REWARD FUNCTION: lender to borrower flow ---- *)
rewardFunction[abilityCombination_, actionCombination_, playerKey_]:=0;

(* SANCTIONS FUNCTION: borrower to deadweight flow ---- *)
sanctionFunction[abilityCombination_, actionCombination_, playerKey_]:=Refine[Piecewise[{
{-Sn,     MemberQ[Delete[abilityCombination,playerKey],0] (* peer cannot repay *)
&&-abilityCombination[[playerKey]]>R (* borrower has more resources than he needs for his own *)
&&-Plus@@(fnGetActionRepayment/@actionCombination)<n R (* borrower does not help *)
}, (* not-helping condition *)
{0,True}
}]];

(* ASSEMBLING PAYOFFLAYERS (apart from repayment) ---- *)
payoffLayers[abilityCombination_, actionCombination_, playerNr_]:=+ punishmentFunction[abilityCombination ,actionCombination, playerNr] + rewardFunction[abilityCombination ,actionCombination, playerNr]+
        sanctionFunction[abilityCombination ,actionCombination, playerNr];

(* ASSUMPTIONS ---- *)
$Assumptions=True; (* Hygiene: Clearing list of global assumptions first . *)
$Assumptions=And[
R> 0, (* sign gross interest *)
Sn>(n-1)R,(* intensity not-helping sanctions *)
M>R]; (* intensity punishment *)
(* Use == for comparisons! No contradictory assumptions! No ambiguous assumptions like <=! *)
```

INDIVIDUAL LIABILITY

(with social sanctions)
Medium punishment intensity
($R < M$)

Punishment strategy

	P2:0R	P2:1R	P2:2R
P1:0R	-M | -M	-M | 0	0 | 0
P1:1R	0 | -M	0 | 0	0 | 0
P1:2R	0 | 0	0 | 0	0 | 0

Ai:	ability to repay for borrower *i*
Pi:	action of player *i*
×	infeasible actions

Map of equilibria

	A2:0	A2:-R	A2:-2 R
A1:0	0R | 0R	0R | 1R	0R | 2R
A1:-R	1R | 0R	1R | 1R	1R | 1R
A1:-2 R	2R | 0R	1R | 1R	1R | 1R

Map of lender payoffs

	A2:0	A2:-R	A2:-2 R
A1:0	0	R	2R
A1:-R	R	2R	2R
A1:-2 R	2R	2R	2R

Map of deadweight loss

	A2:0	A2:-R	A2:-2 R
A1:0	-2M	-M	0
A1:-R	-M	0	0
A1:-2 R	0	0	0

Traceable solution

		A2:0			A2:-R			A2:-2 R		
		P2:0R	P2:1R	P2:2R	P2:0R	P2:1R	P2:2R	P2:0R	P2:1R	P2:2R
A1:0	P1:0R	×[-M,-M]	×	×	×[-M,-M]	×[-M,-R]	×	×[-M,-M-Sn]	×[-M,-R-Sn]	×[0,-2 R]
	P1:1R	×	×	×	×	×	×	×	×	×
	P1:2R	×	×	×	×	×	×	×	×	×
A1:-R	P1:0R	×[-M,-M]	×	×	×[-M,-M]	×[-M,-R]	×	×[-M,-M]	×[-M,-R]	×[0,-2 R]
	P1:1R	×[-R,-M]	×	×	×[-R,-M]	×[-R,-R]	×	×[-R,-M]	×[-R,-R]	×[-R,-2 R]
	P1:2R	×	×	×	×	×	×	×	×	×
A1:-2 R	P1:0R	×[-M-Sn,-M]	×	×	×[-M,-M]	×[-M,-R]	×	×[-M,-M]	×[-M,-R]	×[0,-2 R]
	P1:1R	×[-R-Sn,-M]	×	×	×[-R,-M]	×[-R,-R]	×	×[-R,-M]	×[-R,-R]	×[-R,-2 R]
	P1:2R	×[-2 R,0]	×	×	×[-2 R,0]	×[-2 R,-R]	×	×[-2 R,0]	×[-2 R,-R]	×[-2 R,-2 R]

B.9 Cross-reporting in [RS13] with original defect

```
(* GAME DEFINITION – cross–reporting in RS13 with original defect *)

(* NUMBER OF PLAYERS ---- *)
n=2;

(* ACTIONSET for each player ---- *)
actionSet [1]={
{"0R–n",0,"n"},
{"0R–y",0,"y"},
{"1R",–R,"y"},
{"2R",–2R,"y"}};
actionSet [2]=actionSet [1];

(* PUNISHMENT FUNCTION: borrower to deadweight, controlled by lender ---- *)
punishmentFunction[ abilityCombination_ , actionCombination_, playerKey_]:=Refine[Piecewise[{
(* Rule (1) *)
{0,   (fnGetActionMessage/@actionCombination)==ConstantArray["y",n] (* all say "yes" *)
   && –Plus@@(fnGetActionRepayment/@actionCombination)>= n R (* loan is repaid *)},
(* Rule (2a) *)
{0,   (fnGetActionRepayment/@actionCombination)==ConstantArray[0,n] (* no repayments *)
   && fnGetActionMessage[actionCombination[[playerKey]]]=="n" (* player said "no" *)},
(* Rule (2b) *)
{–P, (fnGetActionRepayment/@actionCombination)==ConstantArray[0,n] (* no repayments *)
   &&fnGetActionMessage[actionCombination[[playerKey]]]=="y"(* player said "yes" *) },
(* Rule (3a) *)
{–P, fnGetActionMessage[actionCombination[[playerKey]]]=="n" (* player said "no" *)
   &&–Plus@@(fnGetActionRepayment/@actionCombination)>= n R (* but loan is repaid *)},
(* Rule (3b) *)
{0, Union[Delete[(fnGetActionMessage/@actionCombination),playerKey]]=={"n"} (* others say no *)
   &&Union[Delete[(fnGetActionRepayment/@actionCombination),playerKey]]=={0} (* others pay zero *)
   &&–fnGetActionRepayment[actionCombination[[playerKey]]]== n R (* player pays all alone *)},
(* Rule (4) *)
{–P, True}
}]];

(* REWARD FUNCTION: lender to borrower flow ---- *)
rewardFunction[ abilityCombination_ , actionCombination_, playerKey_]:=Refine[Piecewise[{
(* Rule (3b) *)
{+n R+e, Union[Delete[(fnGetActionMessage/@actionCombination),playerKey]]=={"n"} (*others say no
   *)
   &&Union[Delete[(fnGetActionRepayment/@actionCombination),playerKey]]=={0} (* others pay zero *)
   &&–fnGetActionRepayment[actionCombination[[playerKey]]]== n R (* player pays all alone *)},
{0,   True}
}]];

(* SANCTIONS FUNCTION: borrower to deadweight flow ---- *)
sanctionFunction [ abilityCombination_ , actionCombination_, playerNr_ ]:=0;

(* ASSEMBLING PAYOFFLAYERS (apart from repayment) ---- *)
payoffLayers [ abilityCombination_ , actionCombination_, playerNr_]:=+ punishmentFunction[
      abilityCombination ,actionCombination , playerNr ] + rewardFunction[ abilityCombination ,
      actionCombination , playerNr]+ sanctionFunction [ abilityCombination ,actionCombination ,
      playerNr ];

(* ASSUMPTIONS ---- *)
$Assumptions=True; (* Hygiene: Clearing list of global assumptions first . *)
$Assumptions=And[
R> 0, (* sign gross interest *)
P>2R, (* intensity punishment *)
e>0]; (* intensity reward *)
(* Use == for comparisons! No contradictory assumptions! No ambiguous assumptions like <=! *)
```

JOINT LIABILITY

(with cross-reporting as in [RS13]) High punishment intensity
($2R < P, 0 < e$)

Punishment and reward strategy

	P2:0R-n	P2:0R-y	P2:1R	P2:2R
P1:0R-n	0 | 0	0 | -P	-P | -P	-P | e+2R
P1:0R-y	-P | 0	-P | -P	-P | -P	0 | 0
P1:1R	-P | -P	-P | -P	0 | 0	0 | 0
P1:2R	e+2R | -P	0 | 0	0 | 0	0 | 0

A*i*:	ability to repay for borrower *i*
P*i*:	action of player *i*
×	infeasible actions

Map of equilibria

	A2:0	A2:-R	A2:-2 R
A1:0	0R-n | 0R-n	0R-n | 0R-n	
A1:-R	0R-n | 0R-n	0R-n | 0R-n 1R | 1R	1R | 1R
A1:-2 R		1R | 1R	1R | 1R

Map of lender payoffs

	A2:0	A2:-R	A2:-2 R
A1:0	0	0	
A1:-R	0	0 2R	2R
A1:-2 R		2R	2R

Map of deadweight loss

	A2:0	A2:-R	A2:-2 R
A1:0	0	0	
A1:-R	0	0 0	0
A1:-2 R		0	0

Traceable solution

A2:0

		P2:0R-n	P2:0R-y	P2:1R	P2:2R
A1:0	P1:0R-n	[0,0]	[0,-P]	×	×
	P1:0R-y	[-P,0]	[-P,-P]	×	×
	P1:1R	×	×	×	×
	P1:2R	×	×	×	×
A1:-R	P1:0R-n	[0,0]	[0,-P]	×	×
	P1:0R-y	[-P,0]	[-P,-P]	×	×
	P1:1R	[-P-R,-P]	[-P-R,-P]	×	×
	P1:2R	×	×	×	×
A1:-2 R	P1:0R-n	[0,0]	[0,-P]	×	×
	P1:0R-y	[-P,0]	[-P,-P]	×	×
	P1:1R	[-P-R,-P]	[-P-R,-P]	×	×
	P1:2R	[e,-P]	[-2 R,0]	×	×

A2:-R

		P2:0R-n	P2:0R-y	P2:1R	P2:2R
A1:0	P1:0R-n	[0,0]	[0,-P]	[-P,-P-R]	×
	P1:0R-y	[-P,0]	[-P,-P]	[-P,-P-R]	×
	P1:1R	×	×	×	×
	P1:2R	×	×	×	×
A1:-R	P1:0R-n	[0,0]	[0,-P]	[-P,-P-R]	×
	P1:0R-y	[-P,0]	[-P,-P]	[-P,-P-R]	×
	P1:1R	[-P-R,-P]	[-P-R,-P]	[-R,-R]	×
	P1:2R	×	×	×	×
A1:-2 R	P1:0R-n	[0,0]	[0,-P]	[-P,-P-R]	×
	P1:0R-y	[-P,0]	[-P,-P]	[-P,-P-R]	×
	P1:1R	[-P-R,-P]	[-P-R,-P]	[-R,-R]	×
	P1:2R	[e,-P]	[-2 R,0]	[-2 R,-R]	×

A2:-2 R

		P2:0R-n	P2:0R-y	P2:1R	P2:2R
A1:0	P1:0R-n	[0,0]	[0,-P]	[-P,-P-R]	[-P,e]
	P1:0R-y	[-P,0]	[-P,-P]	[-P,-P-R]	[0,-2 R]
	P1:1R	×	×	×	×
	P1:2R	×	×	×	×
A1:-R	P1:0R-n	[0,0]	[0,-P]	[-P,-P-R]	[-P,e]
	P1:0R-y	[-P,0]	[-P,-P]	[-P,-P-R]	[0,-2 R]
	P1:1R	[-P-R,-P]	[-P-R,-P]	[-R,-R]	[-R,-2 R]
	P1:2R	×	×	×	×
A1:-2 R	P1:0R-n	[0,0]	[0,-P]	[-P,-P-R]	[-P,e]
	P1:0R-y	[-P,0]	[-P,-P]	[-P,-P-R]	[0,-2 R]
	P1:1R	[-P-R,-P]	[-P-R,-P]	[-R,-R]	[-R,-2 R]
	P1:2R	[e,-P]	[-2 R,0]	[-2 R,-R]	[-2 R,-2 R]

B.10 Cross-reporting in [RS13] with corrected defect

```
(* GAME DEFINITION – cross–reporting in RS13 with corrected defect *)

(* NUMBER OF PLAYERS ---- *)
n=2;

(* ACTIONSET for each player ---- *)
actionSet[1]={
{"0R–n",0,"n"},
{"0R–y",0,"y"},
{"1R",–R,"y"},
{"2R",–2R,"y"}};
actionSet[2]=actionSet[1];

(* PUNISHMENT FUNCTION: borrower to deadweight, controlled by lender ---- *)
punishmentFunction[ abilityCombination_ , actionCombination_, playerKey_]:=Refine[Piecewise[{
(* Rule (1) *)
{0,   (fnGetActionMessage/@actionCombination)==ConstantArray["y",n] (* all say "yes" *)
   && –Plus@@(fnGetActionRepayment/@actionCombination)>= n R (* loan is repaid *)},
(* Rule (2a) *)
{0,   (fnGetActionRepayment/@actionCombination)==ConstantArray[0,n] (* no repayments *)
   && fnGetActionMessage/@actionCombination== ConstantArray["n",n] (* all say "no" *)},
(* Rule (2b) *) (* CHANGED RULE *)
{–P, (fnGetActionRepayment/@actionCombination)==ConstantArray[0,n] (* no repayments *)
   && fnGetActionMessage/@actionCombination!=ConstantArray["n",n] (* not all say "no" *)},
(* Rule (3a) *)
{–P, fnGetActionMessage[actionCombination[[playerKey]]]=="n" (* player said "no" *)
   &&–Plus@@(fnGetActionRepayment/@actionCombination)>= n R (* but loan is repaid *)},
(* Rule (3b) *)
{0,   Union[Delete[(fnGetActionMessage/@actionCombination),playerKey]]=={"n"} (*others say no*)
   &&Union[Delete[(fnGetActionRepayment/@actionCombination),playerKey]]=={0} (*others repay zero
      *)
   &&–fnGetActionRepayment[actionCombination[[playerKey]]]== n R (* player pays all alone *)},
(* Rule (4) *)
{–P, True}
}]];

(* REWARD FUNCTION: lender to borrower flow ---- *)
(* Rule (3b) *)
rewardFunction[ abilityCombination_ , actionCombination_, playerKey_]:=Refine[Piecewise[{
{+n R+e,  Union[Delete[(fnGetActionMessage/@actionCombination),playerKey]]=={"n"} (*others say no
      *)
   &&Union[Delete[(fnGetActionRepayment/@actionCombination),playerKey]]=={0} (*others repay zero
      *)
   &&–fnGetActionRepayment[actionCombination[[playerKey]]]== n R (* player pays all alone *)},
{0, True}
}]];

(* SANCTIONS FUNCTION: borrower to deadweight flow ---- *)
sanctionFunction [ abilityCombination_ , actionCombination_, playerNr_ ]:=0;

(* ASSEMBLING PAYOFFLAYERS (apart from repayment) ---- *)
payoffLayers [ abilityCombination_ , actionCombination_, playerNr_]:=+ punishmentFunction[
      abilityCombination ,actionCombination , playerNr] + rewardFunction[ abilityCombination ,
      actionCombination , playerNr]+ sanctionFunction [ abilityCombination ,actionCombination ,
      playerNr ];

(* ASSUMPTIONS ---- *)
$Assumptions=True; (* Hygiene: Clearing list of global assumptions first . *)
$Assumptions=And[
R> 0, (* sign gross interest *)
P>2R, (* intensity punishment *)
e>0]; (* intensity reward *)
(* Use == for comparisons! No contradictory assumptions! No ambiguous assumptions like <= ! *)
```

INDIVIDUAL LIABILITY

(cross-reporting as in [RS13], corrected)
High punishment intensity ($2R < P, 0 < e$)

Punishment strategy

	P2:0R-n	P2:0R-y	P2:1R	P2:2R
P1:0R-n	0 | 0	-P | -P	-P | -P	-P | e+2R
P1:0R-y	-P | -P	-P | -P	-P | -P	0 | 0
P1:1R	-P | -P	-P | -P	0 | 0	0 | 0
P1:2R	e+2R | -P	0 | 0	0 | 0	0 | 0

Ai:	ability to repay for borrower *i*
Pi:	action of player *i*
×	infeasible actions

Map of equilibria

	A2:0	A2:-R	A2:-2 R
A1:0	0R-n | 0R-n 0R-y | 0R-y	0R-n | 0R-n 0R-y | 0R-y	0R-y | 2R
A1:-R	0R-n | 0R-n 0R-y | 0R-y	0R-n | 0R-n 0R-y | 0R-y 1R | 1R	0R-y | 2R 1R | 1R
A1:-2 R	2R | 0R-y	1R | 1R 2R | 0R-y	0R-y | 2R 1R | 1R 2R | 0R-y

Map of lender payoffs

	A2:0	A2:-R	A2:-2 R
A1:0	0 0	0 0	2R
A1:-R	0 0	0 0 2R	2R 2R
A1:-2 R	2R	2R 2R	2R 2R 2R

Map of deadweight loss

	A2:0	A2:-R	A2:-2 R
A1:0	0 -2P	0 -2P	0
A1:-R	0 -2P	0 -2P 0	0 0
A1:-2 R	0	0 0	0 0 0

Traceable solution

A2:0

		P2:0R-n	P2:0R-y	P2:1R	P2:2R
A1:0	P1:0R-n	[0,0]	[-P,-P]	×	×
	P1:0R-y	[-P,-P]	[-P,-P]	×	×
	P1:1R	×	×	×	×
	P1:2R	×	×	×	×
A1:-R	P1:0R-n	[0,0]	[-P,-P]	×	×
	P1:0R-y	[-P,-P]	[-P,-P]	×	×
	P1:1R	[-P-R,-P]	[-P-R,-P]	×	×
	P1:2R	×	×	×	×
A1:-2 R	P1:0R-n	[0,0]	[-P,-P]	×	×
	P1:0R-y	[-P,-P]	[-P,-P]	×	×
	P1:1R	[-P-R,-P]	[-P-R,-P]	×	×
	P1:2R	[e,-P]	[-2 R,0]	×	×

A2:-R

		P2:0R-n	P2:0R-y	P2:1R	P2:2R
A1:0	P1:0R-n	[0,0]	[-P,-P]	[-P,-P-R]	×
	P1:0R-y	[-P,-P]	[-P,-P]	[-P,-P-R]	×
	P1:1R	×	×	×	×
	P1:2R	×	×	×	×
A1:-R	P1:0R-n	[0,0]	[-P,-P]	[-P,-P-R]	×
	P1:0R-y	[-P,-P]	[-P,-P]	[-P,-P-R]	×
	P1:1R	[-P-R,-P]	[-P-R,-P]	[-R,-R]	×
	P1:2R	×	×	×	×
A1:-2 R	P1:0R-n	[0,0]	[-P,-P]	[-P,-P-R]	×
	P1:0R-y	[-P,-P]	[-P,-P]	[-P,-P-R]	×
	P1:1R	[-P-R,-P]	[-P-R,-P]	[-R,-R]	×
	P1:2R	[e,-P]	[-2 R,0]	[-2 R,-R]	×

A2:-2 R

		P2:0R-n	P2:0R-y	P2:1R	P2:2R
A1:0	P1:0R-n	[0,0]	[-P,-P]	[-P,-P-R]	[-P,e]
	P1:0R-y	[-P,-P]	[-P,-P]	[-P,-P-R]	[0,-2 R]
	P1:1R	×	×	×	×
	P1:2R	×	×	×	×
A1:-R	P1:0R-n	[0,0]	[-P,-P]	[-P,-P-R]	[-P,e]
	P1:0R-y	[-P,-P]	[-P,-P]	[-P,-P-R]	[0,-2 R]
	P1:1R	[-P-R,-P]	[-P-R,-P]	[-R,-R]	[-R,-2 R]
	P1:2R	×	×	×	×
A1:-2 R	P1:0R-n	[0,0]	[-P,-P]	[-P,-P-R]	[-P,e]
	P1:0R-y	[-P,-P]	[-P,-P]	[-P,-P-R]	[0,-2 R]
	P1:1R	[-P-R,-P]	[-P-R,-P]	[-R,-R]	[-R,-2 R]
	P1:2R	[e,-P]	[-2 R,0]	[-2 R,-R]	[-2 R,-2 R]

B.11 Cross-reporting in [RS04]

```
(* GAME DEFINITION -- cross-reporting as in RS04 *)

(* Auxilary functions to transform an actionCombination into an actionKeyCombination *)
fnActionToKey[playerKey_, action_]:=Sequence@@@Position[actionSet[playerKey],action ];
fnActionToKeyCombination[actionCombination_]:=Flatten[MapIndexed[fnActionToKey[Sequence@@
        #2,#1]&,actionCombination,{1}]];

(* case in the paper where h>=2R and thus A=min[h,2R]=2R *)
A=2R

(* NUMBER OF PLAYERS ---- *)
n=2;

(* ACTIONSET for each player ---- *)
actionSet[1]={
{"0R-n",0,"n"},
{"0R-y",0,"y"},
{"1R",-R,""},
{"2R",-2R,""}};
actionSet[2]=actionSet[1];

(* PUNISHMENT FUNCTION: borrower to deadweight, controlled by lender ---- *)
punishmentRule[1]={
{-R, -(A+e), -A, -P },
{-P, -P,      -P, -(2R-A) },
{ 0,  -P,      0, 0 },
{ 0,   0,     -P, 0 }};
punishmentRule[2]=Transpose[punishmentRule[1]];

(* Punishment function chooses the adequate entries from the matrices defined above *)
punishmentFunction[ abilityCombination_ , actionCombination_, playerKey_]:=punishmentRule[playerKey ][[
        Sequence@@fnActionToKeyCombination[actionCombination]]];

(* REWARD FUNCTION: lender to borrower flow ---- *)
rewardRule[1]={
{0, 0,  0,  0},
{0, 0,  0,  0},
{0, 0,  0,  R+e},
{A-R+e, 0, 0,  0}};
rewardRule[2]=Transpose[rewardRule[1]];

(* Reward function chooses the adequate entries from the matrices defined above *)
rewardFunction[ abilityCombination_ , actionCombination_, playerKey_]:=rewardRule[playerKey ][[
        Sequence@@fnActionToKeyCombination[actionCombination]]];

(* SANCTIONS FUNCTION: borrower to deadweight flow ---- *)
sanctionFunction [ abilityCombination_ , actionCombination_, playerNr_ ]:=0;

(* ASSEMBLING PAYOFFLAYERS (apart from repayment) ---- *)
payoffLayers [ abilityCombination_ , actionCombination_, playerNr_]:=+ punishmentFunction[
        abilityCombination ,actionCombination,playerNr] + rewardFunction[ abilityCombination ,
        actionCombination,playerNr]+ sanctionFunction [ abilityCombination ,actionCombination,
        playerNr ];

(* ASSUMPTIONS ---- *)
$Assumptions=True; (* Hygiene: Clearing list of global assumptions first . *)
$Assumptions=And[
R> 0, (* sign gross interest *)
P>2R, (* intensity punishment *)
0<e<P-A]; (* intensity reward *)
(* Use == for comparisons! No contradictory assumptions! No ambiguous assumptions like <=! *)
```

JOINT LIABILITY

(with cross-reporting as in [RS04])
High punishment intensity
$(2R < P, 0 < e < P - 2R)$

Punishment and reward strategy

	P2:0R-n	P2:0R-y	P2:1R	P2:2R
P1:0R-n	-R | -R	-e-2R | -P	-2R | 0	-P | e+R
P1:0R-y	-P | -e-2R	-P | -P	-P | -P	0 | 0
P1:1R	0 | -2R	-P | -P	0 | 0	e+R | -P
P1:2R	e+R | -P	0 | 0	-P | e+R	0 | 0

Symbol in [RS04]	Here
h	none
$\Psi > h$	P
$\bar{R}$	R
$A \equiv \min\{h, 2\bar{R}\}$	$2R$
$0 < \varepsilon < \Psi - h$	e
$(0, \mu)$	"0R-y"

Map of equilibria

	A2:0	A2:-R	A2:-2 R
A1:0	0R-n | 0R-n	0R-n | 0R-n 0R-n | 1R	0R-y | 2R
A1:-R	0R-n | 0R-n 1R | 0R-n	0R-n | 0R-n 1R | 1R	1R | 1R
A1:-2 R	2R | 0R-y	1R | 1R	1R | 1R

Map of lender payoffs

	A2:0	A2:-R	A2:-2 R
A1:0	0	0 R	2R
A1:-R	0 R	0 2R	2R
A1:-2 R	2R	2R	2R

Map of deadweight loss

	A2:0	A2:-R	A2:-2 R
A1:0	-2R	-2R -2R	0
A1:-R	-2R -2R	-2R 0	0
A1:-2 R	0	0	0

Traceable solution

A2:0

		P2:0R-n	P2:0R-y	P2:1R	P2:2R
A1:0	P1:0R-n	[-R,-R]	[-e-2 R,-P]	×	×
	P1:0R-y	[-P,-e-2 R]	[-P,-P]	×	×
	P1:1R	×	×	×	×
	P1:2R	×	×	×	×
A1:-R	P1:0R-n	[-R,-R]	[-e-2 R,-P]	×	×
	P1:0R-y	[-P,-e-2 R]	[-P,-P]	×	×
	P1:1R	[-R,-2 R]	[-P-R,-P]	×	×
	P1:2R	×	×	×	×
A1:-2 R	P1:0R-n	[-R,-R]	[-e-2 R,-P]	×	×
	P1:0R-y	[-P,-e-2 R]	[-P,-P]	×	×
	P1:1R	[-R,-2 R]	[-P-R,-P]	×	×
	P1:2R	[e-R,-P]	[-2 R,0]	×	×

A2:-R

		P2:0R-n	P2:0R-y	P2:1R	P2:2R
A1:0	P1:0R-n	[-R,-R]	[-e-2 R,-P]	[-2 R,-R]	×
	P1:0R-y	[-P,-e-2 R]	[-P,-P]	[-P,-P-R]	×
	P1:1R	×	×	×	×
	P1:2R	×	×	×	×
A1:-R	P1:0R-n	[-R,-R]	[-e-2 R,-P]	[-2 R,-R]	×
	P1:0R-y	[-P,-e-2 R]	[-P,-P]	[-P,-P-R]	×
	P1:1R	[-R,-2 R]	[-P-R,-P]	[-R,-R]	×
	P1:2R	×	×	×	×
A1:-2 R	P1:0R-n	[-R,-R]	[-e-2 R,-P]	[-2 R,-R]	×
	P1:0R-y	[-P,-e-2 R]	[-P,-P]	[-P,-P-R]	×
	P1:1R	[-R,-2 R]	[-P-R,-P]	[-R,-R]	×
	P1:2R	[e-R,-P]	[-2 R,0]	[-P-2 R,e]	×

A2:-2 R

		P2:0R-n	P2:0R-y	P2:1R	P2:2R
A1:0	P1:0R-n	[-R,-R]	[-e-2 R,-P]	[-2 R,-R]	[-P,e-R]
	P1:0R-y	[-P,-e-2 R]	[-P,-P]	[-P,-P-R]	[0,-2 R]
	P1:1R	×	×	×	×
	P1:2R	×	×	×	×
A1:-R	P1:0R-n	[-R,-R]	[-e-2 R,-P]	[-2 R,-R]	[-P,e-R]
	P1:0R-y	[-P,-e-2 R]	[-P,-P]	[-P,-P-R]	[0,-2 R]
	P1:1R	[-R,-2 R]	[-P-R,-P]	[-R,-R]	[e,-P-2 R]
	P1:2R	×	×	×	×
A1:-2 R	P1:0R-n	[-R,-R]	[-e-2 R,-P]	[-2 R,-R]	[-P,e-R]
	P1:0R-y	[-P,-e-2 R]	[-P,-P]	[-P,-P-R]	[0,-2 R]
	P1:1R	[-R,-2 R]	[-P-R,-P]	[-R,-R]	[e,-P-2 R]
	P1:2R	[e-R,-P]	[-2 R,0]	[-P-2 R,e]	[-2 R,-2 R]

B.12 Cross-reporting in [RS04] improved

```
(* GAME DEFINITION - cross-reporting as in RS04 *)

(* Auxilary functions to transform an actionCombination into an actionKeyCombination *)
fnActionToKey[playerKey_, action_]:=Sequence@@@Position[actionSet[playerKey],action ];
fnActionToKeyCombination[actionCombination_]:=Flatten[MapIndexed[fnActionToKey[Sequence@@
        #2,#1]&,actionCombination, {1}]];

(* case in the paper where h>=2R and thus A=min[h,2R]=2R *)
A=2R

(* NUMBER OF PLAYERS ---- *)
n=2;

(* ACTIONSET for each player ---- *)
actionSet [1]={
{"0R-n",0,"n"},
{"0R-y",0,"y"},
{"1R",-R,""},
{"2R",-2R,""}};
actionSet [2]= actionSet [1];

(* PUNISHMENT FUNCTION: borrower to deadweight, controlled by lender ---- *)
punishmentRule[1]={
{-M, -(A+e), -M, -P },
{-P, -P,      -P, -(2R-A) },
{ 0,  -P,      0, 0 },
{ 0,   0,     -P, 0 }};
punishmentRule[2]=Transpose[punishmentRule[1]];

(* Punishment function chooses the adequate entries from the matrices defined above *)
punishmentFunction[ abilityCombination_ , actionCombination_, playerKey_]:=punishmentRule[playerKey
        ][[ Sequence@@fnActionToKeyCombination[actionCombination]]];

(* REWARD FUNCTION: lender to borrower flow ---- *)
rewardRule[1]={
{0, 0,  0,  0},
{0, 0,  0,  0},
{0, 0,  0,  R+e},
{A-R+e, 0, 0,  0}};
rewardRule[2]=Transpose[rewardRule[1]];

(* Reward function chooses the adequate entries from the matrices defined above *)
rewardFunction[ abilityCombination_ , actionCombination_, playerKey_]:=rewardRule[playerKey ][[
        Sequence@@fnActionToKeyCombination[actionCombination]]];

(* SANCTIONS FUNCTION: borrower to deadweight flow ---- *)
sanctionFunction [ abilityCombination_ , actionCombination_, playerNr_ ]:=0;

(* ASSEMBLING PAYOFFLAYERS (apart from repayment) ---- *)
payoffLayers [ abilityCombination_ , actionCombination_, playerNr_]:=+ punishmentFunction[
        abilityCombination , actionCombination , playerNr ] + rewardFunction[ abilityCombination ,
        actionCombination , playerNr]+ sanctionFunction [ abilityCombination , actionCombination ,
        playerNr ];

(* ASSUMPTIONS ---- *)
$Assumptions=True; (* Hygiene: Clearing list of global assumptions first . *)
$Assumptions=And[
R> 0, (* sign gross interest *)
P>2R, (* intensity high punishment *)
R<M<2R, (* intensity medium punishment *)
0<e<P-A]; (* intensity reward *)
(* Use ==for comparisons! No contradictory assumptions! No ambiguous assumptions like <=! *)
```

JOINT LIABILITY

(cross-reporting [RS04] improved)
High punishment intensity
$(2R < P, 0 < e < P - 2R, R < M < 2R)$

Punishment and reward strategy

	P2:0R-n	P2:0R-y	P2:1R	P2:2R
P1:0R-n	-M | -M	-e-2R | -P	-M | 0	-P | e+R
P1:0R-y	-P | -e-2R	-P | -P	-P | -P	0 | 0
P1:1R	0 | -M	-P | -P	0 | 0	e+R | -P
P1:2R	e+R | -P	0 | 0	-P | e+R	0 | 0

A*i*:	ability to repay for borrower *i*
P*i*:	action of player *i*
×	infeasible actions

Map of equilibria

	A2:0	A2:-R	A2:-2 R
A1:0	0R-n | 0R-n	0R-n | 1R	0R-y | 2R
A1:-R	1R | 0R-n	1R | 1R	1R | 1R
A1:-2 R	2R | 0R-y	1R | 1R	1R | 1R

Map of lender payoffs

	A2:0	A2:-R	A2:-2 R
A1:0	0	R	2R
A1:-R	R	2R	2R
A1:-2 R	2R	2R	2R

Map of deadweight loss

	A2:0	A2:-R	A2:-2 R
A1:0	-2M	-M	0
A1:-R	-M	0	0
A1:-2 R	0	0	0

Traceable solution

A2:0

		P2:0R-n	P2:0R-y	P2:1R	P2:2R
A1:0	P1:0R-n	×[-M,-M]	×[-e-2 R,-P]	×	×
	P1:0R-y	×[-P,-e-2 R]	×[-P,-P]	×	×
	P1:1R	×	×	×	×
	P1:2R	×	×	×	×
A1:-R	P1:0R-n	×[-M,-M]	×[-e-2 R,-P]	×	×
	P1:0R-y	×[-P,-e-2 R]	×[-P,-P]	×	×
	P1:1R	×[-R,-M]	×[-P-R,-P]	×	×
	P1:2R	×	×	×	×
A1:-2 R	P1:0R-n	×[-M,-M]	×[-e-2 R,-P]	×	×
	P1:0R-y	×[-P,-e-2 R]	×[-P,-P]	×	×
	P1:1R	×[-R,-M]	×[-P-R,-P]	×	×
	P1:2R	×[e-R,-P]	×[-2 R,0]	×	×

A2:-R

		P2:0R-n	P2:0R-y	P2:1R	P2:2R
A1:0	P1:0R-n	×[-M,-M]	×[-e-2 R,-P]	×[-M,-R]	×
	P1:0R-y	×[-P,-e-2 R]	×[-P,-P]	×[-P,-P-R]	×
	P1:1R	×	×	×	×
	P1:2R	×	×	×	×
A1:-R	P1:0R-n	×[-M,-M]	×[-e-2 R,-P]	×[-M,-R]	×
	P1:0R-y	×[-P,-e-2 R]	×[-P,-P]	×[-P,-P-R]	×
	P1:1R	×[-R,-M]	×[-P-R,-P]	×[-R,-R]	×
	P1:2R	×	×	×	×
A1:-2 R	P1:0R-n	×[-M,-M]	×[-e-2 R,-P]	×[-M,-R]	×
	P1:0R-y	×[-P,-e-2 R]	×[-P,-P]	×[-P,-P-R]	×
	P1:1R	×[-R,-M]	×[-P-R,-P]	×[-R,-R]	×
	P1:2R	×[e-R,-P]	×[-2 R,0]	×[-P-2 R,e]	×

A2:-2 R

		P2:0R-n	P2:0R-y	P2:1R	P2:2R
A1:0	P1:0R-n	×[-M,-M]	×[-e-2 R,-P]	×[-M,-R]	×[-P,e-R]
	P1:0R-y	×[-P,-e-2 R]	×[-P,-P]	×[-P,-P-R]	×[0,-2 R]
	P1:1R	×	×	×	×
	P1:2R	×	×	×	×
A1:-R	P1:0R-n	×[-M,-M]	×[-e-2 R,-P]	×[-M,-R]	×[-P,e-R]
	P1:0R-y	×[-P,-e-2 R]	×[-P,-P]	×[-P,-P-R]	×[0,-2 R]
	P1:1R	×[-R,-M]	×[-P-R,-P]	×[-R,-R]	×[e,-P-2 R]
	P1:2R	×	×	×	×
A1:-2 R	P1:0R-n	×[-M,-M]	×[-e-2 R,-P]	×[-M,-R]	×[-P,e-R]
	P1:0R-y	×[-P,-e-2 R]	×[-P,-P]	×[-P,-P-R]	×[0,-2 R]
	P1:1R	×[-R,-M]	×[-P-R,-P]	×[-R,-R]	×[e,-P-2 R]
	P1:2R	×[e-R,-P]	×[-2 R,0]	×[-P-2 R,e]	×[-2 R,-2 R]

B.13 Cross-reporting with individual liability

```
(* GAME DEFINITION - cross-reporting with individual liability *)

(* NUMBER OF PLAYERS ---- *)
n=2;

(* ACTIONSET for each player ---- *)
actionSet [1]={
{"0R-n",0,"n"},
{"0R-y",0,"y"},
{"1R-n",-R,"n"},
{"1R-y",-R,"y"},
{"2R-n",-2R,"n"},
{"2R-y",-2R,"y"}};
actionSet [2]= actionSet [1];

(* PUNISHMENT FUNCTION: borrower to deadweight, controlled by lender ---- *)
punishmentFunction[ abilityCombination_ , actionCombination_, playerKey_]:=Refine[Piecewise[{
(* anti-coordination-failure-rule *)
{-2M, (fnGetActionMessage/@actionCombination)==ConstantArray["y",n](* if both say "yes" *)
&& -Plus@@(fnGetActionRepayment/@actionCombination)==0 (* and nobody repays *)},
(* anti-lying-rule one *)
{-M, -Plus@@Delete[(fnGetActionRepayment/@actionCombination),playerKey]>=R(* if peer repays *)
&& fnGetActionMessage[actionCombination[[playerKey]]]=="n" (* but player said "no" *)},
(* anti-lying-rule two *)
{-M, -Plus@@Delete[(fnGetActionRepayment/@actionCombination),playerKey]==0(* if peer defaults *)
&& fnGetActionMessage[actionCombination[[playerKey]]]=="y" (* but player said "yes" *)},
(* anti-strategic-default-rule *)
{-M, Delete[(fnGetActionMessage/@actionCombination),playerKey]=={"y"}(* if peer "yes" *)
&& fnGetActionRepayment[actionCombination[[playerKey]]]==0 (* but player defaults *)},
{0, True}
}]];

(* REWARD FUNCTION: lender to borrower flow ---- *)
rewardFunction[ abilityCombination_ , actionCombination_, playerKey_]:=Refine[Piecewise[{
(* sincerity-reward-rule *)
{+e+R, Delete[(fnGetActionMessage/@actionCombination), playerKey]== ConstantArray["n",n-1] (*
        peer says "no" *)
&&-fnGetActionRepayment[actionCombination[[playerKey]]]>=R (*player repays despite peer message*)
        },
{0,True}
}]];

(* SANCTIONS FUNCTION: borrower to deadweight flow ---- *)
sanctionFunction [ abilityCombination_ , actionCombination_, playerNr_ ]:=0;

(* ASSEMBLING PAYOFFLAYERS (apart from repayment) ---- *)
payoffLayers [ abilityCombination_ , actionCombination_, playerNr_]:=+ punishmentFunction[
        abilityCombination ,actionCombination ,playerNr] + rewardFunction[ abilityCombination ,
        actionCombination ,playerNr]+ sanctionFunction [ abilityCombination ,actionCombination,
        playerNr ];

(* ASSUMPTIONS ---- *)
$Assumptions=True; (* Hygiene: Clearing list of global assumptions first . *)
$Assumptions=And[
R> 0, (* sign gross interest *)
M>R, (* intensity punishment *)
e>0(* intensity reward *)
];
(* Use == for comparisons! No contradictory assumptions! No ambiguous assumptions like <=! *)
```

INDIVIDUAL LIABILITY

(with cross-messaging)
Medium punishment intensity
$(R < M, 0 < e)$

Punishment and reward strategy

	P2:0R-n	P2:0R-y	P2:1R-n	P2:1R-y	P2:2R-n	P2:2R-y
P1:0R-n	0 | 0	-M | -M	-M | e+R	-M | e-M+R	-M | e+R	-M | e-M+R
P1:0R-y	-M | -M	-2M | -2M	0 | 0	-M | -M	0 | 0	-M | -M
P1:1R-n	e+R | -M	0 | 0	e-M+R | e-M+R	-M | e+R	e-M+R | e-M+R	-M | e+R
P1:1R-y	e-M+R | -M	-M | -M	e+R | -M	0 | 0	e+R | -M	0 | 0
P1:2R-n	e+R | -M	0 | 0	e-M+R | e-M+R	-M | e+R	e-M+R | e-M+R	-M | e+R
P1:2R-y	e-M+R | -M	-M | -M	e+R | -M	0 | 0	e+R | -M	0 | 0

Map of equilibria

	A2:0	A2:-R	A2:-2 R
A1:0	0R-n | 0R-n	0R-y | 1R-n	0R-y | 1R-n
A1:-R	1R-n | 0R-y	1R-y | 1R-y	1R-y | 1R-y
A1:-2 R	1R-n | 0R-y	1R-y | 1R-y	1R-y | 1R-y

Map of lender payoffs

	A2:0	A2:-R	A2:-2 R
A1:0	0	R	R
A1:-R	R	2R	2R
A1:-2 R	R	2R	2R

Map of deadweight loss

	A2:0	A2:-R	A2:-2 R
A1:0	0	0	0
A1:-R	0	0	0
A1:-2 R	0	0	0

Traceable solution

A2:0

		P2:0R-n	P2:0R-y	P2:1R-n	P2:1R-y	P2:2R-n	P2:2R-y
A1:0	P1:0R-n	[0,0]	[-M,-M]	×	×	×	×
	P1:0R-y	[-M,-M]	[-2 M,-2 M]	×	×	×	×
	P1:1R-n	×	×	×	×	×	×
	P1:1R-y	×	×	×	×	×	×
	P1:2R-n	×	×	×	×	×	×
	P1:2R-y	×	×	×	×	×	×
A1:-R	P1:0R-n	[0,0]	[-M,-M]	×	×	×	×
	P1:0R-y	[-M,-M]	[-2 M,-2 M]	×	×	×	×
	P1:1R-n	[e,-M]	[-R,0]	×	×	×	×
	P1:1R-y	[e-M,-M]	[-M-R,-M]	×	×	×	×
	P1:2R-n	×	×	×	×	×	×
	P1:2R-y	×	×	×	×	×	×
A1:-2 R	P1:0R-n	[0,0]	[-M,-M]	×	×	×	×
	P1:0R-y	[-M,-M]	[-2 M,-2 M]	×	×	×	×
	P1:1R-n	[e,-M]	[-R,0]	×	×	×	×
	P1:1R-y	[e-M,-M]	[-M-R,-M]	×	×	×	×
	P1:2R-n	[e-R,-M]	[-2 R,0]	×	×	×	×
	P1:2R-y	[e-M-R,-M]	[-M-2 R,-M]	×	×	×	×

A2:-R

		P2:0R-n	P2:0R-y	P2:1R-n	P2:1R-y	P2:2R-n	P2:2R-y
A1:0	P1:0R-n	[0,0]	[-M,-M]	[-M,e]	[-M,e-M]	×	×
	P1:0R-y	[-M,-M]	[-2 M,-2 M]	[0,-R]	[-M,-M-R]	×	×
	P1:1R-n	×	×	×	×	×	×
	P1:1R-y	×	×	×	×	×	×
	P1:2R-n	×	×	×	×	×	×
	P1:2R-y	×	×	×	×	×	×
A1:-R	P1:0R-n	[0,0]	[-M,-M]	[-M,e]	[-M,e-M]	×	×
	P1:0R-y	[-M,-M]	[-2 M,-2 M]	[0,-R]	[-M,-M-R]	×	×
	P1:1R-n	[e,-M]	[-R,0]	[e-M,e-M]	[-M-R,e]	×	×
	P1:1R-y	[e-M,-M]	[-M-R,-M]	[e,-M-R]	[-R,-R]	×	×
	P1:2R-n	×	×	×	×	×	×
	P1:2R-y	×	×	×	×	×	×
A1:-2 R	P1:0R-n	[0,0]	[-M,-M]	[-M,e]	[-M,e-M]	×	×
	P1:0R-y	[-M,-M]	[-2 M,-2 M]	[0,-R]	[-M,-M-R]	×	×
	P1:1R-n	[e,-M]	[-R,0]	[e-M,e-M]	[-M-R,e]	×	×
	P1:1R-y	[e-M,-M]	[-M-R,-M]	[e,-M-R]	[-R,-R]	×	×
	P1:2R-n	[e-R,-M]	[-2 R,0]	[e-M-R,e-M]	[-M-2 R,e]	×	×
	P1:2R-y	[e-M-R,-M]	[-M-2 R,-M]	[e-R,-M-R]	[-2 R,-R]	×	×

	A2:-2 R						
		P2:0R-n	P2:0R-y	P2:1R-n	P2:1R-y	P2:2R-n	P2:2R-y
A1:0	P1:0R-n	[0,0]	[-M,-M]	[-M,e]	[-M,e-M]	[-M,e-R]	[-M,e-M-R]
	P1:0R-y	[-M,-M]	[-2 M,-2 M]	[0,-R]	[-M,-M-R]	[0,-2 R]	[-M,-M-2 R]
	P1:1R-n	×	×	×	×	×	×
	P1:1R-y	×	×	×	×	×	×
	P1:2R-n	×	×	×	×	×	×
	P1:2R-y	×	×	×	×	×	×
		P2:0R-n	P2:0R-y	P2:1R-n	P2:1R-y	P2:2R-n	P2:2R-y
A1:-R	P1:0R-n	[0,0]	[-M,-M]	[-M,e]	[-M,e-M]	[-M,e-R]	[-M,e-M-R]
	P1:0R-y	[-M,-M]	[-2 M,-2 M]	[0,-R]	[-M,-M-R]	[0,-2 R]	[-M,-M-2 R]
	P1:1R-n	[e,-M]	[-R,0]	[e-M,e-M]	[-M-R,e]	[e-M,e-M-R]	[-M-R,e-R]
	P1:1R-y	[e-M,-M]	[-M-R,-M]	[e,-M-R]	[-R,-R]	[e,-M-2 R]	[-R,-2 R]
	P1:2R-n	×	×	×	×	×	×
	P1:2R-y	×	×	×	×	×	×
		P2:0R-n	P2:0R-y	P2:1R-n	P2:1R-y	P2:2R-n	P2:2R-y
A1:-2 R	P1:0R-n	[0,0]	[-M,-M]	[-M,e]	[-M,e-M]	[-M,e-R]	[-M,e-M-R]
	P1:0R-y	[-M,-M]	[-2 M,-2 M]	[0,-R]	[-M,-M-R]	[0,-2 R]	[-M,-M-2 R]
	P1:1R-n	[e,-M]	[-R,0]	[e-M,e-M]	[-M-R,e]	[e-M,e-M-R]	[-M-R,e-R]
	P1:1R-y	[e-M,-M]	[-M-R,-M]	[e,-M-R]	[-R,-R]	[e,-M-2 R]	[-R,-2 R]
	P1:2R-n	[e-R,-M]	[-2 R,0]	[e-M-R,e-M]	[-M-2 R,e]	[e-M-R,e-M-R]	[-M-2 R,e-R]
	P1:2R-y	[e-M-R,-M]	[-M-2 R,-M]	[e-R,-M-R]	[-2 R,-R]	[e-R,-M-2 R]	[-2 R,-2 R]

Bibliography

[AB13] Lutz Arnold and Benedikt Booker. Good intentions pave the way to … the local moneylender. *Economics Letters*, 118(3):466–469, 2013.

[AG00] Beatriz Armendáriz and Christian Gollier. Peer group formation in an adverse selection model. *Economic Journal*, 110(465):632–643, 2000.

[Agu09] Itai Agur. On the possibility of credit rationing in the Stiglitz-Weiss model: a comment. Working Paper 237, De Nederlandsche Bank, December 2009. Available at: `http://citeseerx.ist.psu.edu/viewdoc/download?doi=10.1.1.456.5885&rep=rep1&type=pdf`.

[AH00] Reinnhold Adrian and Thomas Heidorn. *Der Bankbetrieb*. Gabler Verlag, fifteenth edition, August 2000.

[Ahl15] Christian Ahlin. The role of group size in group lending. *Journal of Development Economics*, 115:140–155, 2015.

[Ake70] George Arthur Akerlof. The market for "lemons": Quality Uncertainty and the Market Mechanism. *Quarterly Journal of Economics*, 84(3):488–500, August 1970.

[All16] Treb Allen. Optimal (partial) group liability in microfinance lending. *Journal of Development Economics*, 121:201–216, 2016.

[AM10] Beatriz Armendáriz and Jonathan Morduch. *The Economics of Microfinance*. MIT Press, second edition, 2010.

[Ame16] American Law Institute. *What Can We Learn from Credit Markets*, 2016. 93rd Annual Meeting, Washington, D.C., May 18, 2016. Available at: `https://www.ali.org/media/filer_public/41/2f/412f21ed-e4a3-49c6-aadc-8c7cca0a750f/simkovic-remarks.pdf`.

[Ani03] Kumar Aniket. Sequential group lending with moral hazard. Working Paper. Available at: `www.researchgate.net`, June 2003.

C. Prem, *The Theory of Credit Contracts*,
https://doi.org/10.1007/978-3-658-29362-8

[AR09] Lutz Arnold and John Riley. On the possibility of credit rationing in the Stiglitz-Weiss model. *American Economic Review*, pages 2012–2021, 2009.

[Arm99] Beatriz Armendáriz. On the design of a credit agreement with peer monitoring. *Journal of Development Economics*, 60(1):79–104, 1999.

[Arm11] Beatriz Armendáriz. Women and microsavings. In *The Handbook of Microfinance*, pages 503–516. World Scientific Publishing Co. Pte. Ltd., 2011.

[Arn16] Lutz Arnold. *Makroökonomik*. Mohr Siebeck Tübingen, fifth edition, 2016.

[ARS13] Lutz G. Arnold, Johannes Reeder, and Susanne Steger. On the viability of group lending when microfinance meets the market: A reconsideration of the besley—coate model. *Journal of Emerging Market Finance*, 12(1):59–106, 2013.

[ASP15] Ashta Arvind, Khan Saleh, and Otto Philipp. Does Microfinance Cause or Reduce Suicides? Policy Recommendations for Reducing Borrower Stress. *Strategic Change*, 24(2):165–190, March 2015.

[AT07] Christian Ahlin and Robert M. Townsend. Selection into and across credit contracts: Theory and field research. *Journal of Econometrics*, 136(2):665–698, February 2007.

[Bal78] Ernst Baltensperger. Credit Rationing: Issues and Questions. *Journal of Money, Credit and Banking*, 10(2):170–183, 1978.

[Bar76] Robert J. Barro. The loan market, collateral, and rates of interest. *Journal of Money, Credit and Banking*, 8(4):439–456, November 1976.

[BC95] Timothy Besley and Stephen Coate. Group lending, repayment incentives and social collateral. *Journal of Development Economics*, 46(1):1–18, 1995.

[BD05] Patrick Bolton and Mathias Dewatripont. *Contract Theory*. MIT Press, 2005.

[BDN05] John H. Boyd and Gianni De Nicoló. The Theory of Bank Risk Taking and Competition Revisited. *The Journal of Finance*, 60(3):1329–1343, June 2005.

[Bes85] Helmut Bester. Screening vs. rationing in credit markets with imperfect information. *The American Economic Review*, 75(4):850–855, 1985.

[Bes87] Helmut Bester. The role of collateral in credit markets with imperfect information. *European Economic Review*, 31(4):887–899, 1987.

[BK93] David Besanko and George Kanatas. Credit Market Equilibrium with Bank Monitoring and Moral Hazard. *The Review of Financial Studies*, 6(1):213–232, 1993.

[BK04] Philip Bond and Arvind Krishnamurthy. Regulating exclusion from financial markets. *The Review of Economic Studies*, 71(3):681–707, 2004.

[BKS14] Allen N. Berger, Thomas Kick, and Klaus Schaeck. Executive board composition and bank risk taking. *Journal of Corporate Finance*, 28:48–65, 2014.

[BO99] Tamer Basar and Geert Jan Olsder. *Dynamic Noncooperative Game Theory*. Society for Industrial and Applied Mathematics, second edition, 1999.

[BO10] Bharat Bhole and Sean Ogden. Group lending and individual lending with strategic default. *Journal of Development Economics*, 91:348–363, 2010.

[BR02] Philip Bond and Ashok Rai. Collateral substitutes in microfinance. Working Paper. Available at: `http://finance.wharton.upenn.edu/~{}pbond/research/limits-jul2-02.pdf`, July 2002.

[BS90] Patrick Bolton and David S. Scharfstein. A Theory of Predation Based on Agency Problems in Financial Contracting. *American Economic Review*, 80(1):93–106, March 1990.

[BS94] John H. Boyd and Bruce D. Smith. How Good Are Standard Debt Contracts? Stochastic Versus Nonstochastic Monitoring in a Costly State Verification Environment. *Journal of Business*, 67(4):539–561, October 1994.

[BS08] Dyuti Banerjee and Anupama Sethi. Intra-Group Transfers And Group Formation. Working paper, Monash University, Department of Economics, 2008. Available at: `https://www.monash.edu/business/economics/research/publications/2008/2408intragroupbanerjeesethi.pdf`.

[BS12] Sylvain Bourjade and Ibolya Schindele. Group lending with endogenous group size. *Economics Letters*, 117(3):556–560, 2012.

[BSW14] Jean-Marie Baland, Rohini Somanathan, and Zaki Wahhaj. Group lending and endogenous social sanctions. Discussion Paper No. 1415, University of Kent, School of Economics, 2014. Available at: `https://www.econstor.eu/bitstream/10419/129983/1/814788696.pdf`.

[BU90] Allen N. Berger and Gregory F. Udell. Collateral, loan quality and bank risk. *Journal of Monetary Economics*, 25(1):21–42, 1990.

[BZ06] Giovanni Busetta and Alberto Zazzaro. Mutual Loan-Guarantee Societies in Credit Markets with Adverse Selection: Do They Act as a Sorting Device? Working Paper. Available at: `www.researchgate.net`, 2006.

[Cho05] Prabal Roy Chowdhury. Group-lending: Sequential financing, lender monitoring and joint liability. *Journal of Development Economics*, 77(2):415–439, August 2005.

[Cho07] Prabal Roy Chowdhury. Group-lending with sequential financing, contingent renewal and social capital. *Journal of Development Economics*, 84(1):487–506, 2007.

[Chu11] Craig Churchill. Insurance for the Poor: Definitions and Innovations. In *The Handbook of Microfinance*, pages 537–562. World Scientific Publishing, 2011.

[CK85] Yuk-Shee Chan and George Kanatas. Asymmetric Valuations and the Role of Collateral in Loan Agreements. *Journal of Money, Credit and Banking*, 17(1):84–95, February 1985.

[Coc99] Giuseppe Coco. Collateral, heterogeneity in risk attitude and the credit market equilibrium. *European Economic Review*, 43(3):559–574, 1999.

[Coc00] Giuseppe Coco. On the use of collateral. *Journal of Economic Surveys*, 14(2):191–214, 2000.

[Con99] Jonathan Conning. Outreach, sustainability and leverage in monitored and peer-monitored lending. *Journal of Development Economics*, 60(1):51–77, 1999.

[Con05] Jonathan Conning. Monitoring by Delegates or by Peers? Joint Liability Loans under Moral Hazard. Working Paper. Available at:

http://econ.hunter.cuny.edu/wp-content/uploads/sites/6/RePEc/papers/HunterEconWP407.pdf, June 2005.

[CS05] Prabirendra Chatterjee and Sudipta Sarangi. Enforcement with Costly Group Formation. *Economics Bulletin*, 15(9):1–8, February 2005.

[DD05] Jean-Pierre Danthine and John Donaldson. *Intermediate Financial Theory*. Elsevier Academic Press, second edition, 2005.

[Dia84] Douglas W. Diamond. Financial Intermediation and Delegated Monitoring. *Review of Economic Studies*, 51(3):393–414, 1984.

[DMW87] David De Meza and David C. Webb. Too Much Investment: A Problem of Asymmetric Information. *Quarterly Journal of Economics*, 102(2):281–292, 1987.

[DNDLV10] Gianni De Nicolò, Giovanni Dell'Ariccia, Luc Laeven, and Fabian Valencia. Monetary Policy and Bank Risk Taking. IMF Staff Position Note SPN/10/09, International Monetary Fund, July 2010. Available at: http://citeseerx.ist.psu.edu/viewdoc/download?doi=10.1.1.172.813&rep=rep1&type=pdf.

[Dos16] Anastasios Dosis. On Signalling and Screening in Markets with Asymmetric Information. Working Paper 1608, ESSEC Business School, 2016. Available at: https://hal-essec.archives-ouvertes.fr/hal-01285190v2/document.

[DQFG16] Jonathan De Quidt, Thiemo Fetzer, and Maitreesh Ghatak. Group lending without joint liability. *Journal of Development Economics*, 121:217–236, 2016.

[DR94] Prajit K. Dutta and Roy Radner. Moral Hazard. In R. J. Aumann and S. Hart, editors, *Handbook of Game Theory*, volume 2, chapter 26, pages 869–903. Elsevier Science B. V., 1994.

[DS63] John M. Dutton and William H. Starbuck. On Managers and Theories. *Management International*, 3(6):25–35, 1963.

[EMS02] Paul Embrechts, Alexander McNeil, and Daniel Straumann. Correlation and Dependence in Risk Management: Properties and Pitfalls. In M. A. H. Dempster, editor, *Risk Management: Value at Risk and Beyond*, pages 176–223. Cambridge University Press, 2002.

[Fei59] Hermann Feifel. *Die Anwendbarkeit der modernen Kreditschöpfungslehre auf die besondere Art des Sparkassengeschäfts*. Duncker & Humblot, Berlin, 1959.

[FG10] Greg Fischer and Maitreesh Ghatak. Repayment Frequency in Microfinance Contracts with Present-Biased Borrowers. Working paper, London School of Economics, July 2010. Available at: `http://eprints.lse.ac.uk/58184/1/__lse.ac.uk_storage_LIBRARY_Secondary_libfile_shared_repository_Content_STICERD_EOPP_eopp21.pdf`.

[FG11] Greg Fischer and Maitreesh Ghatak. Spanning the Chasm: Uniting Theory and Empirics in Microfinance Research. In *The Handbook of Microfinance*, chapter 3, pages 59–75. World Scientific Publishing, 2011.

[FP08] Erica Field and Rohini Pande. Repayment Frequency and Default in Microfinance: Evidence from India. *Journal of the European Economic Association*, 6(2-3):501–509, May 2008.

[FR08] Xavier Freixas and Jean-Charles Rochet. *Microeconomics of Banking*. MIT Press, second edition, 2008.

[FR15] Robert Frontczak and Stefan Rostek. Modeling loss given default with stochastic collateral. *Economic Modelling*, 44:162–170, January 2015.

[GG99] Maitreesh Ghatak and Timothy W. Guinnane. The economics of lending with joint liability: theory and practice. *Journal of Development Economics*, 60:195–228, 1999.

[GGH+13] Ludwig Gramlich, Peter Gluchowski, Andreas Horsch, Klaus Schäfer, and Gerd Waschbusch. *Gabler Bank-Lexikon*. Springer Gabler, fourteenth edition, 2013.

[GH85] Douglas Gale and Martin Hellwig. Incentive-Compatible Debt Contracts: The One-Period Problem. *Review of Economic Studies*, 52(4):647–663, 1985.

[Gha99] Maitreesh Ghatak. Group lending, local information and peer selection. *Journal of Development Economics*, 60(1):27–50, 1999.

[Gha00] Maitreesh Ghatak. Screening by the company you keep: Joint liability lending and the peer selection effect. *Economic Journal*, 110(465):601–631, July 2000.

[GK14] Xavier Giné and Dean S. Karlan. Group versus individual liability: Short and long term evidence from Philippine microcredit lending groups. *Journal of Development Economics*, 107:65–83, 2014.

[GL14] Shubhashis Gangopadhyay and Robert Lensink. Asymmetric group loans, non-assortative matching and adverse selection. *Economics Letters*, 124(2):185–187, 2014.

[GMRS11] Isabelle Guérin, Solène Morvant-Roux, and Jean-Michel Servet. Understanding the Diversity and Complexity of Demand for Microfinance Services: Lessons from Informal Finance. In *The Handbook of Microfinance*, pages 101–122. World Scientific Publishing, 2011.

[Gro94] Denis Gromb. Renegotiation in debt contracts. Working Paper. Available at: `http://citeseerx.ist.psu.edu/viewdoc/download?doi=10.1.1.512.4264&rep=rep1&type=pdf`, May 1994.

[Gut08] Joel M. Guttman. Assortative matching, adverse selection, and group lending. *Journal of Development Economics*, 87(1):51–56, 2008.

[GW11] Christophe J. Godlewski and Laurent Weill. Does Collateral Help Mitigate Adverse Selection? A Cross-Country Analysis. *Journal of Financial Services Research*, 40(1):49–78, October 2011.

[Han01] Frank Hanser. *Die Struktur von Kreditbeziehungen*. Gabler Verlag und Deutscher Universitäts-Verlag, first edition, 2001.

[HCG06] Valentina Hartarska, Steven B. Caudill, and Daniel M. Gropper. The Cost Structure of Microfinance Institutions in Eastern Europe and Central Asia. Working Paper Number 809, William Davidson Institute at the University of Michigan, January 2006. Available at: `https://deepblue.lib.umich.edu/bitstream/handle/2027.42/40195/wp809.pdf?sequence=3`.

[HL07] Niels Hermes and Robert Lensink. Impact of Microfinance: A Critical Survey. *Economic and Political Weekly*, 42(6):462–465, February 2007.

[Hol79] Bengt Holmström. Moral Hazard and Observability. *Bell Journal of Economics*, 10(1):74–91, 1979.

[Hol82] Bengt Holmström. Moral Hazard in Teams. *Bell Journal of Economics*, 13(2):324–340, 1982.

[HT97] Bengt Holmstrom and Jean Tirole. Financial intermediation, loanable funds, and the real sector. *Quarterly Journal of Economics*, 112(3):663–691, 1997.

[Imb05] Kathryn Imboden. Building inclusive financial sectors: The road to growth and poverty reduction. *Journal of International Affairs*, 58(2):65–86, 2005.

[Imp98] Gregorio Impavido. Credit rationing, group lending and optimal group size. *Annals of Public and Cooperative Economics*, 69(2):243–260, 1998.

[Inn90] Robert D. Innes. Limited liability and incentive contracting with ex-ante action choices. *Journal of Economic Theory*, 52(1):45–67, 1990.

[Jea03] Tirole Jean. Incomplete Contracts: Where Do We Stand? *Econometrica*, 67(4):741–781, December 2003.

[Jen09] Mark William Jenkins. *Essays on Consumer Credit Markets*. phdthesis, Department of Economics at Stanford University, December 2009.

[JN02] Robert A. Jones and David Nickerson. Mortgage Contracts, Strategic Options and Stochastic Collateral. *Journal of Real Estate Finance and Economics*, 24(1):35–58, Jan 2002.

[JPB05] Tullio Jappelli, Marco Pagano, and Magda Bianco. Courts and Banks: Effects of Judicial Enforcement on Credit Markets. *Journal of Money, Credit and Banking*, 37(2):223–244, April 2005.

[Kar07] Dean S. Karlan. Social Connections and Group Banking. *Economic Journal*, 117:F52–F84, February 2007.

[Kee79] William R. Keeton. *Equilibrium Credit Rationing*. Garland Publishing, 1979.

[Key36] John Maynard Keynes. *The General Theory of Employment, Interest and Money*. Macmillan, London, 1936.

[KL12] Tomek Katzur and Robert Lensink. Group lending with correlated project outcomes. *Economics Letters*, 117(2):445–447, 2012.

[KOU14] Kei Kawai, Ken Onishi, and Kosuke Uetake. Signaling in Online Credit Markets. Working paper. Available at: `https://vlab.decisionsciences.columbia.edu/newsletter/Mar0413/signaling%20in%20online.pdf`, August 2014.

[KP98] Fahad Khalil and Bruno M. Parigi. Loan size as a commitment device. *International Economic Review*, 39(1):135–150, February 1998.

[KS94] David M. Kreps and Joel Sobel. Signalling. In R. J. Aumann and S. Hart, editors, *Handbook of Game Theory*, volume 2, chapter 25, pages 849–867. Elsevier Science, 1994.

[KS14] Artashes Karapetyan and Bogdan Stacescu. Does information sharing reduce the role of collateral as a screening device? *Journal of Banking and Finance*, 43:48–57, 2014.

[KV00] Stefan Krasa and Anne P. Villamil. Optimal Contracts When Enforcement is a Decision Variable. *Econometrica*, 68(1):119–134, January 2000.

[KV11] Dean Karlan and Martin Valdivia. Teaching Entrepreneurship: Impact of Business Training on Microfinance Clients and Institutions. *Review of Economics and Statistics*, 93(2):510–527, May 2011.

[Laf03] Jean-Jacques Laffont. Collusion and group lending with adverse selection. *Journal of Development Economics*, 70(2):329–348, 2003.

[Led98] Joanna Ledgerwood. *Microfinance Handbook: An Institutional and Financial Perspective*. World Bank Publications, 1998.

[Man28] Thomas Mann. *The Magic Mountain*. Secker & Warburg London, third edition, 1928. Translated from the German, 1971 Reprint.

[MCWG95] Andreu Mas-Colell, Michael D. Whinston, and Jerry R. Green. *Microeconomic Theory*. Oxford University Press, 1995.

[MFE05] Alexander J. McNeil, Rüdiger Frey, and Paul Embrechts. *Quantitative Risk Management: Concepts, Techniques and Tools*. Princeton University Press, first edition, 2005.

[MNS06] Lukas Menkhoff, Doris Neuberger, and Chodechai Suwanaporn. Collateral-based lending in emerging markets: Evidence from Thailand. *Journal of Banking and Finance*, 30(1):1–21, 2006.

[MO14] Sana Mohsni and Isaac Otchere. Risk taking behavior of privatized banks. *Journal of Corporate Finance*, 29:122–142, 2014.

[MP86] Arie Melnik and Steven Plaut. Loan commitment contracts, terms of lending, and credit allocation. *Journal of Finance*, 41(2):425–435, 1986.

[MP89] Dilip Mookherjee and Ivan Png. Optimal Auditing, Insurance, and Redistribution. *Quarterly Journal of Economics*, 104(2):399–415, May 1989.

[MQ02] Michael Magill and Martine Quinzii. *Theory of Incomplete Markets*, volume 1. MIT Press, 2002.

[MR88] Hellmuth Milde and John G. Riley. Signaling in credit markets. *Quarterly Journal of Economics*, 103(1):101–129, 1988.

[MS02] Eric Maskin and Tomas Sjöström. Implementation Theory. In K. J. Arrow, A. K. Sen, and K. Suzumura, editors, *Handbook of Social Choice and Welfare*, volume 1, chapter 5, pages 237–288. Elsevier Science, 2002.

[MS13] R. Mersland and R. Øystein Strøm. Microfinance: Costs, Lending Rates, and Profitability. In Gerard Caprio, editor, *Handbook of Key Global Financial Markets, Institutions, and Infrastructure*, chapter 44, pages 489–499. Elsevier, first edition, 2013.

[MW13] Sebastian C. Moenninghoff and Axel Wieandt. The Future of Peer-to-Peer Finance. *Zeitschrift für Betriebswirtschaftliche Forschung*, 65:466–487, September 2013.

[Nel06] Roger B. Nelsen. *An Introduction to Copulas*. Springer, New York, second edition, 2006.

[Nii09] Juha-Pekka Niinimäki. Screening in the credit market when the collateral value is stochastic. Discussion Paper 19, Bank of Finland Research, 2009. Available at: `https://helda.helsinki.fi/bof/bitstream/handle/123456789/7830/163910.pdf`.

[Nii11] Juha-Pekka Niinimäki. Nominal and true cost of loan collateral. *Journal of Banking and Finance*, 35(10):2782–2790, October 2011.

[NS05] Noam Nisan and Shimon Schocken. *The Elements of Computing Systems*. The MIT Press, 2005. Paperback edition, 2008.

[Par00] Cheol Park. Monitoring and structure of debt contracts. *Journal of Finance*, 55(5):2157–2195, October 2000.

[PHS14] Luminita Postelnicu, Niels Hermes, and Ariane Szafarz. Defining Social Collateral in Microfinance Group Lending. In Roy Mersland and R. Øystein Strøm, editors, *Microfinance Institutions: Financial and Social Performance*, chapter 10, pages 187–207. Palgrave Studies in Impact Finance, first edition, 2014.

[PMF17] Achim Peters, Bruce S. McEwen, and Karl Friston. Uncertainty and stress: Why it causes diseases and how it is mastered by the brain. *Progress in Neurobiology*, 156:164–188, September 2017.

[Rah99] Aminur Rahman. Micro-credit Initiatives for Equitable and Sustainable Development: Who Pays? *World Development*, 27(1):67–82, 1999.

[Ras07] Eric Rasmusen. *Games and Information: An Introduction to Game Theory*. Blackwell Publishing, fourth edition, 2007.

[RC12] David Rowell and Luke B. Connelly. A History of the Term "Moral Hazard". *Journal of Risk and Insurance*, 79(4):1051–1075, 2012.

[Ric12] Stefan Richter. *Schadenszurechnung bei deliktischer Haftung für fehlerhafte Sekundärmarktinformation*. Mohr Siebeck, Tübingen, 2012.

[RS70] Michael Rothschild and Joseph E. Stiglitz. Increasing Risk: I. A Definition. *Journal of Economic Theory*, 2(3):225–243, September 1970.

[RS00] Rafael Repullo and Javier Suarez. Entrepreneurial moral hazard and bank monitoring: A model of the credit channel. *European Economic Review*, 44(10):1931–1950, 2000.

[RS04] Ashok S. Rai and Tomas Sjöström. Is Grameen Lending Efficient? Repayment Incentives and Insurance in Village Economies. *Review of Economic Studies*, 71(1):217–234, 2004.

[RS13] Ashok Rai and Tomas Sjöström. Redesigning Microcredit. In Nir Vulkan, Alvin E. Roth, and Zvika Neeman, editors, *The Handbook of Market Design*, chapter 9, pages 249–265. Oxford University Press, 2013.

[Rut11] Stuart Rutherford. Boosting the Poor's Capacity to Save: A Note on Instalment Plans and their Variants. In *The Handbook of Microfinance*, pages 517–535. World Scientific Publishing, 2011.

[Sch01] Mark Schreiner. Seven Aspects of Loan Size. *Journal of Microfinance*, 3(2):27–47, 2001.

[Sha03] Tridib Sharma. Optimal Contracts When Enforcement Is a Decision Variable: A Comment. *Econometrica*, 71(1):387–390, January 2003.

[Sha12] Andrey Shatelyuk. *Sanierungskredite in der Krise und in der Insolvenz von Unternehmen: Eine vergleichende Untersuchung des deutschen und russischen Rechts*. De Gruyter, 2012.

[Sim56] Herbert A. Simon. Rational Choice and the Structure of the Environment. *Psychological Review*, 63(2):129–138, 1956.

[Ste10] Susanne E. Steger. *Kreditmarktunvollkommenheiten und Mikrokredite*. PhD thesis, University of Regensburg, 2010.

[Ste11] Christian Stegbauer. *Reziprozität - Einführung in soziale Formen der Gegenseitigkeit*. Verlag für Sozialwissenschaften, second edition, 2011.

[Sti90] Joseph E. Stiglitz. Peer Monitoring and Credit Markets. *World Bank Economic Review*, 4(3):351–366, September 1990.

[SW81] Joseph E. Stiglitz and Andrew Weiss. Credit Rationing in Markets with Imperfect Information. *American Economic Review*, 71(3):393–410, 1981.

[SW06] Sam Schulhofer-Wohl. Negative assortative matching of risk-averse agents with transferable expected utility. *Economics Letters*, 92(3):383–388, 2006.

[SZ06] Franklin Simtowe and Manfred Zeller. Determinants of Moral Hazard in Microfinance: Empirical Evidence from Joint Liability Lending Programs in Malawi. Working paper, Munich Personal RePEc Archive, October 2006. Available at: `https://mpra.ub.uni-muenchen.de/461/1/MPRA_paper_461.pdf`.

[Sza97] Nick Szabo. Formalizing and Securing Relationships on Public Networks. *First Monday*, 2(9), September 1997. Available at: `http://firstmonday.org/ojs/index.php/fm/article/view/548`.

[Tow79] Robert M. Townsend. Optimal Contracts and Competitive Markets with Costly State Verification. *Journal of Economic Theory*, 21:265–293, 1979.

[Wei03] Karl E. Weick. Theory and Practice in the Real World. In Haridimos Tsoukas and Christian Knudsen, editors, *The Oxford Handbook of Organization Theory*, chapter 16, pages 453–475. Oxford University Press, 2003.

[Zel98] Manfred Zeller. Determinants of Repayment Performance in Credit Groups: The Role of Program Design, Intragroup Risk Pooling, and

Social Cohesion. *Economic Development and Cultural Change*, 46(3):599–620, April 1998.

[Zha08] Wei Zhang. Ex-ante Moral Hazard and Repayment Performance under Group Lending. *Journal of East Asian Studies*, 6:145–171, 2008.